Biological Science

Control and co-ordination in organisms
a Laboratory guide

Edited by John A. Barker

14 082603 3

Nuffield Advanced Science

Advanced Biological Science

Organizers — P. J. Kelly, Centre for Science Education, Chelsea College, University of London

W. H. Dowdeswell, School of Education, University of Bath: formerly Winchester College

Development team — J. A. Barker, Centre for Science Education, Chelsea College, University of London: formerly Borough Road College

B. J. Bull

L. C. Comber

J. F. Eggleston, School of Education, University of Leicester

W. H. Freeman, Chislehurst and Sidcup Grammar School for Boys

P. J. Fry, Brentwood College of Education: formerly Westcliff High School for Boys

J. H. Gray, Department of Education, University of Keele: formerly Manchester Grammar School

R. E. Lister, Institute of Education, University of London: formerly Wolverhampton Grammar School

B. S. Mowl, Centre for Science Education, Chelsea College, University of London: formerly Churchfields Comprehensive School

M. K. Sands, Department of Education, University of Nottingham: formerly King Edward VI Camp Hill School for Girls, Birmingham

C. F. Stoneman, Department of Education, University of York: formerly Dulwich College

B. J. K. Tricker, Victoria College, Jersey: formerly Eton College

K. O. Turner, Department of Education, University of Cambridge: formerly Hinckley Grammar School

Biological Science

Control and co-ordination in organisms
in organisms
a Laboratory guide

Edited by John A. Barker

Nuffield Advanced Science
Published for the Nuffield Foundation by Penguin Books

Penguin Books Ltd, Harmondsworth, Middlesex, England
Penguin Books Inc., 7110 Ambassador Road,
Baltimore, Md 21207, U.S.A.
Penguin Books Ltd, Ringwood, Victoria, Australia

Copyright © The Nuffield Foundation, 1970

Design and art direction by Ivan and Robin Dodd
Illustrations designed and produced by Penguin Education

Filmset in Monophoto Ionic
by Oliver Burridge Filmsetting Ltd, Crawley
and made and printed in Great Britain
by Fleming & Humphreys (Baylis) Ltd, Leicester and London

Contents

Foreword Sixth form courses in Britain have received more than their fair share of blessing and cursing in the last twenty years: blessing, because their demands, their compass, and their teachers are often of a standard which in other countries would be found in the first year of a longer university course than ours: cursing, because this same fact sets a heavy cloud of university expectation on their horizon (with awkward results for those who finish their education at the age of 18) and limits severely the number of subjects that can be studied in the sixth form.

So advanced work, suitable for students between the ages of 16 and 18, is at the centre of discussions on the curriculum. It need not, of course, be in a 'sixth form' at all, but in an educational institution other than a school. In any case, the emphasis on the requirements of those who will not go to a university or other institute of higher education is increasing, and will probably continue to do so; and the need is for courses which are satisfying and intellectually exciting in themselves – not for courses which are simply passports to further study.

Advanced Science Courses are therefore both an interesting and a difficult venture. Yet fresh work on Advanced science teaching was obviously needed if new approaches to the subject (with all the implications that these have for pupils' interest in learning science and adults' interest in teaching it) were not to fail in their effect. The Trustees of the Nuffield Foundation therefore agreed to support teams, on the same model as had been followed in their other science projects, to produce Advanced courses in Physical Science, in Physics, in Chemistry, and in Biological Science. It was realized that the task would be an immense one, partly because of the universities' special interest in the approach and content of these courses, partly because the growing size of sixth forms underlined the point that Advanced work was not *solely* a preparation for a degree course, and partly because the blending of Physics and Chemistry in a single Advanced Physical Science course was bound to produce problems. Yet, in spite of these pressures, the emphasis here, as in the other Nuffield Science courses, is on learning rather than on being taught, on understanding rather than amassing information, on finding out rather than on being told: this emphasis is central to all worthwhile attempts at curriculum renewal.

If these Advanced courses meet with the success and appreciation which I believe they deserve, then the credit will belong to a large number of people, in the teams and the consultative committees, in schools and universities, in authorities and councils and associations and boards: once again it has been the Foundation's privilege to provide a point at which the imaginative and helpful efforts of many could come together.

Brian Young
Director of the Nuffield Foundation

General Editor of the *Laboratory guides:*
W. H. Dowdeswell

Editor of this volume:
John A. Barker

Contributors:
John A. Barker
W. H. Freeman
C. F. Stoneman

Preface   The purpose of this *Laboratory guide* is to provide a coherent treatment of a series of biological topics, based on laboratory investigations. The approach throughout is centred on enquiry as opposed to the mere verification of facts and through this work you will gain not only knowledge but also an understanding of the processes by which biological knowledge is acquired. Observation, the design of investigations, working out ideas – hypotheses – on which to base experiments, analysing results, and drawing relevant conclusions from them will play a fundamental role in the work you do.

No matter how logically it may appear to hang together, a sequence of laboratory investigations inevitably leaves large and important gaps of knowledge uncovered. For this reason each investigation is preceded by a short *introductory section*, intended to give it relevance and relate it with what has gone before and what is to follow. Following this, detailed instructions are provided under the title *Procedure*. However, you will find that a good deal of latitude has been left for individual initiative and interpretation in designing and carrying out experiments. At the end of each investigation there is a short series of *Questions*. It is advisable to read these carefully before embarking upon the relevant investigation as, to some extent, they point the direction in which the enquiry should be conducted.

At the end of each chapter is a short and carefully selected *Bibliography*. This lists the titles of a few suitable books (it makes no pretence at being exhaustive) which will enable you to broaden your understanding of the topics you have investigated in the laboratory and see how they are related to others. At the same time the list indicates the range of topics that you should aim to cover in your reading.

In the short period of time available it is obviously impossible to span more than a fraction of the subject matter by practical work. So, just as in scientific research, we must acquire our knowledge both by laboratory studies and by using fully the findings and interpretations of others.

Each investigation in the *Laboratory guide* is numbered in the lefthand margin, using a decimal system to allow easy cross reference.

Complementary to the series of *Laboratory guides* is the *Study guide*. This is also a book of investigations but it does not involve practical work and makes use of various types of second hand data instead. It serves also as a further source of theoretical information and is integrated with the work in the *Laboratory guides*.

Synopsis

1 Organisms are unable to survive without an adequate water supply.

2 Survival also depends on an organism's ability to regulate the amount of water it takes in and gives out.

3 The rate of uptake and output of water of a plant is regulated by structures of the plant and factors in the environment.

4 Xylem tissue in flowering plants is suited to the function of water transport.

5 The active response that animals make to maintain themselves in an optimum environment can be investigated with an organism such as *Tribolium*.

Chapter 1

The organism and water

Summary of practical work

section	topic
1.1	Do roots affect the rate at which a plant takes up water?
1.2	Is the rate of water uptake affected by environmental conditions around the aerial portion of the plant?
1.3	What part does the leaf play in determining the amount of water lost by the plant?
1.4	Transport of water in a plant stem
1.5	The anatomy of a stem and how water goes up
1.6	Responses of *Tribolium* to humidity

Water and survival

Two-thirds (63–67 per cent) of the weight of the body of an adult human being is water. The body weight of herbaceous plants and such animals as jellyfish consists of up to 90 per cent or more of water. Indeed, water is an essential constituent of all living organisms and without it they cannot survive. Moreover, a man will die more quickly from thirst than from starvation. Nevertheless, survival does not merely depend upon the presence of water but also on the ability of the organism to regulate the amount of water it takes in and gives out. If a terrestrial plant is left out of water, it becomes limp and droops. This process of wilting will continue and eventually the plant will die. Yet the same plant submerged in water will also die. In contrast to this, consider the case of aquatic plants and animals. They cannot survive for long out of water, or even with small quantities of water, unless they possess special adaptations, for example, the operculum possessed by some aquatic snails by which they can close the opening of their shells. Aquatic plants and animals can, however, withstand being under water, which, to a terrestrial organism, is lethal. Observations such as these indicate that organisms survive only when a suitable amount of water is available.

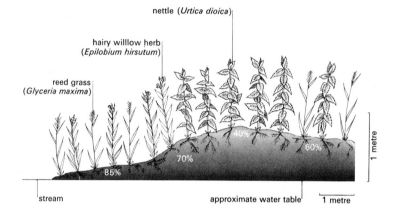

Figure 1
An area of fenland in the Itchen valley, Hampshire, showing the distribution of plants at different heights. The percentages give the content of water in the topsoil. *After Baron, W. M. M. (1967)* Organisation in plants, *Edward Arnold.*

Figure 1 shows a profile through some Hampshire fenland. The mineral content and texture of the soil are similar throughout, but, depending on the height of the ground above the water table, the water content varies; this, apparently, is what determines the distribution of the plants found there. The nettle *(Urtica dioica)* survives best in a soil with 40 per cent water content. The hairy willow herb *(Epilobium hirsutum)* appears most suited to about 70 per cent. Although the reed grass *(Glycera maxima)* has a wide range of tolerance, when it is in competition with the other two species, it grows best in soil with a water content of 85 per cent and above.

Clearly the adaptations possessed by these plants for regulating the amount of water passing in and out of their bodies are important factors enabling them to compete with other species in their environment. What are these adaptations? How is the water content of organisms regulated?

Factors affecting the input and output of water in a plant

Water normally passes into terrestrial plants through the roots and ascends the stem; some escapes by evaporation into the atmosphere through the leaves. This movement of water is called the transpiration stream. We can investigate the conditions under which the processes take place, and the rate at which they occur, by simple experiments using small potted plants or shoots. *Impatiens balsamina* is a suitable plant.

Consider investigations 1.1–1.3 and carry out one or more of the suggested experimental procedures.

Before you start, formulate a hypothesis from the topic of the investigation, which you can test experimentally. Taking the outline procedure as a basis, plan your investigation. Use an adequate experimental technique and, where appropriate, suitable controls.

1.1 Do roots affect the rate at which a plant takes up water ?

Figure 2
Investigating the uptake of water by plants, with and without roots.

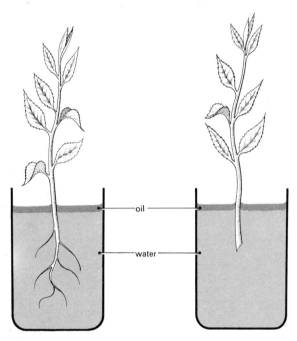

oil

water

Procedure

1   Figure 2 shows a simple experimental technique to investigate the effect of roots on the uptake of water.
2   By weighing the whole apparatus at suitable intervals we can record any loss in weight. For the purpose of this experiment we shall assume that this is due to transpiration.

Questions   a   Does the presence of roots significantly affect the rate of water uptake?

b   Under standard conditions is the rate of loss in weight constant?

c   If we remove a part or all of the leaves, what effect does this have on the rate of loss?

d   How could you establish that the loss in weight was in fact due to transpiration?

1.2 Is the rate of water uptake affected by environmental conditions around the aerial portions of the plant?

Procedure   1   Figure 3 shows a simple potometer which can be used to measure the uptake of water by a cut shoot.
Select a suitable shoot which has a stem circular in cross section. Cut the shoot one or two inches longer than required and immediately plunge the cut end into a suitable container of water.

2   Hold the end under water and cut it a second time, using a slanting cut, so that the end is thick enough to provide a firm fit when inserted into the plastic tubing.

3   Place the plastic tubing with the capillary attached into the container of water. When all the air has been expelled, carefully insert the cut end of the shoot into the plastic tubing. If necessary, gently twist a piece of thin wire around the connection.

4   Set up a suitable stand, beaker, and block as in the illustration. It is best to boil the water beforehand and let it cool, as this will help to prevent bubbles of gas appearing in the tube. Place a finger over the end of the capillary tube, remove the tube and the shoot, and set them up as shown in the illustration. Dry and wet leaves and allow the apparatus to settle down for five minutes.

5   The rate of uptake of water can be measured by introducing a bubble into the end of the capillary tube. To do this remove the block, lower the beaker, and let a small air bubble into the capillary tube. Attach a suitable scale to the capillary tube, and so measure the rate of movement of the bubble. When it reaches the upper end pinch the plastic tube gently to expel the bubble and restart.

6   Take an adequate series of readings under normal conditions and then alter the environmental conditions around the shoot. You could investigate, by suitable techniques, the effects of darkness, movement of air, temperature, and humidity changes.

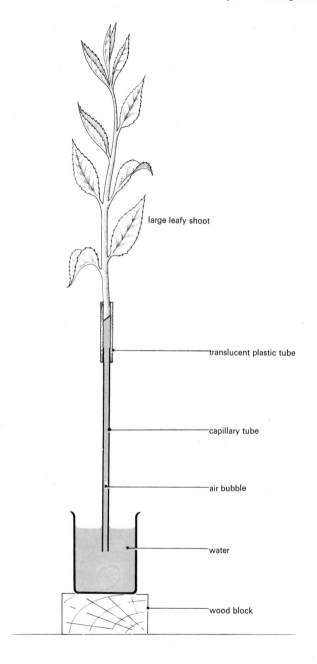

Figure 3
A simple potometer. The uptake of water is measured by the movement of the air bubble along the scale on the capillary tube.

large leafy shoot

translucent plastic tube

capillary tube

air bubble

water

wood block

Questions

a   How much water passed through the shoot under normal laboratory conditions, per unit time, per unit weight?

b   Under what conditions does the shoot take up water
1 most rapidly?
2 least rapidly?

1.3 What part does the leaf play in determining the amount of water lost by the plant?

Procedure

1   Using a potometer, investigate the effect of leaves on the uptake of water by removing these, one by one, from the shoot.

2   Investigate changes in leaf weight with a microbalance (figure 4a). Assuming these to be due to evaporation of water, compare the loss in weight of a leaf with that of a piece of wet paper of the same area over the same period.

3   Compare the rate of loss in weight from leaves of the same type, but differing in area. The technique of measuring the area of a leaf is shown in figure 4b.

4   Smearing petroleum jelly over the surface of leaves will produce a waterproof coat. Using this technique investigate whether transpiration occurs mainly through the upper or lower surfaces of leaves.

5   Take thin strips of epidermis, or replicas, from the upper and lower surfaces of the leaves. Make temporary microscope slide preparations and examine these for stomata.

Questions

a   What part do leaves play in the rate of transpiration of a shoot?

b   Does a leaf work as a simple physical evaporator of water, or is there any evidence that it can control the rate of water loss?

c   Does the rate of transpiration from a leaf depend on its surface area?

d   Is the transpiration from the surface of a leaf linked with the presence or absence of stomata?

Figure 4
Means of measuring the loss in weight
of leaves.

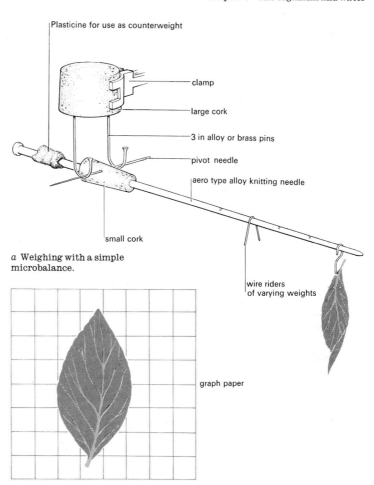

a Weighing with a simple
microbalance.

b Measuring the area before weighing
leaves of different sizes.

1.4 Transport of water in a plant stem

The stem connects the root system of a plant to the leaves
and therefore must contain the means of transport by which
the water, taken in from the soil, reaches the leaves. What
part of the stem is concerned with water transport?

Procedure

1 Cut a vigorous shoot from a plant with a translucent stem,
such as *Impatiens balsamina*.
2 Immediately immerse the cut end in an aqueous solution
of a dye, such as eosin or methylene blue. Support the shoot
in an upright position.

3   Examine the movement of the dye with a hand lens. (The stem should be translucent enough to enable you to do this.)

4   When the dye has reached the leaf tissues, remove the shoot and, using a moistened razor blade, cut transverse sections, not more than 1 mm thick, from the stem. See figure 5.

Figure 5
Cutting a transverse section with a razor. This view is from *above*. The stem is held vertically, the razor horizontally.

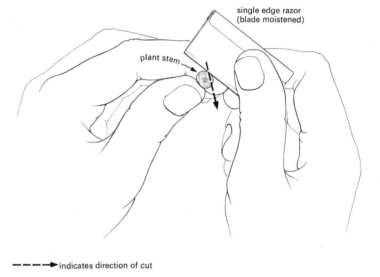

single edge razor (blade moistened)

plant stem

- - - -► indicates direction of cut

Place the sections on a microscope slide, moisten with a drop of water, and examine them, using a lower power objective.

Questions

a   To which specific areas of the stem is the dye confined?

b   What anatomical structures are found in these areas?

c   In your earlier investigations you may have determined some of the conditions under which shoots lose water rapidly. Can you devise and carry out a further simple experiment to find out whether the rate at which a dye solution has passed up the stem of a cut shoot is related to the environmental conditions around it?

1.5 The anatomy of a stem and how water goes up

Procedure

1   Take a length of stem and, holding it as shown in figure 5, cut thin transverse sections off it using a moistened razor-blade. Cut several sections and use the thinnest for further examination. You do not need a complete section across the stem; a small sector will be sufficient. Place the thin section in a little water on a slide.

Procedure

1   Select a suitable position on a bench which is out of direct sunlight and not near a heat source, such as a radiator or bench lamp. Do not jar the bench when you are doing the experiment.

2   Place the choice chamber on a piece of light-proof material, so that half of it can be folded over to cover the chamber.

3   Arrange the chamber so that the end of the central bar, in the lower part, faces you, dividing the chamber into right and left halves. Use drawing pins to fix the chamber to the bench on either side so that it cannot be moved.
    Take the upper part and fix muslin across the open base, using an elastic band. Pull the material taut to remove any folds or creases. Fix the two parts of the chamber securely together.

4   Drop five adult beetles through the left hole and then seal it. Do the same with the right hole. The animals can be collected and transferred either by using a pooter, a device for capturing and transferring small arthropods, or with an artist's small paint brush.

5   Cover the apparatus with light-proof material so that all light is excluded. After 10–15 minutes remove the cover and record the number of animals in each half of the chamber. Recover and repeat your observations after a further 10 minutes, and if possible, after another 10 minutes. Collect data from as many choice chambers as possible in order to analyse them.

6   Remove the upper part containing the animals and place it on a sheet of paper. Fill one section of the lower part with water, taking care not to get water into the other section. Alternatively, soak a piece of plastic sponge, cut to size and fitted into the section, with water. Fill the other section with a drying agent, e.g. calcium chloride. Set up some chambers with the left side, and others with the right, as the wet side. This will act as a control against the effect of other stimuli, such as a sloping bench top, which may influence the behaviour of the animals.

7   Replace the upper part containing the animals. If some cobalt thiocyanate paper is available, you can make a check on the humidity gradient by inserting the thin strip through one hole and laying this over the central bar at right angles. Cobalt thiocyanate paper is blue when in a dry atmosphere and pink in a humid one.

Figure 9
A choice chamber. Attach gauze to
the base of the upper ring.

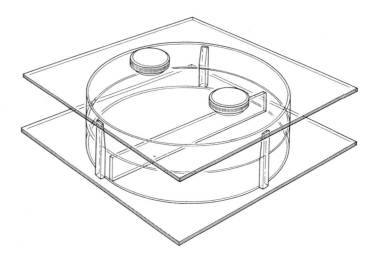

8   Cover the chamber to exclude light as before. Leave it
undisturbed for 15 minutes to 20 minutes to allow the
humidity gradient to be established, and then record the
animals in the two sections. If you have time, remove the
top section, turn it through 180°, and replace it. Leave it
undisturbed, as before, for 15–20 minutes and then record the
animals in the two sections.
Collect data from as many choice chambers as possible in
order that they can be analysed statistically.

Questions

a   What is the hypothesis on which this experiment is based?
b   What was the purpose of the first series of observations on
the beetles (stage 5 above)? What distribution would you
expect in this case?
c   What responses, if any, did the beetles make to a humidity
gradient?
d   Supposing that in some cases a small proportion of the
beetles were found consistently in one section and the larger
proportion in the other. What hypotheses can you suggest
to explain this result? How could you test them?
e   Under natural conditions the beetles would be in darkness.
Perhaps light would affect their orientation. If you have
time, you could set up a further investigation to discover
whether they behave similarly in the light and in the dark.

Bibliography

Baron, W.M.M. (1967) *Water and plant life*. Heinemann Educational Books. (Useful account of the relationship of water and plants. Particularly relevant for the movement of water through the plant.)

Carthy, J.D. (1966) *The study of behaviour*. Edward Arnold. (A follow-up providing further details about orientation mechanisms.)

Richardson, M. (1968) *Translocation in plants*. Edward Arnold. (Clear account of the structure of xylem and the experimental evidence of its transport function.)

Steward, F.C. (1964) *Plants at work*. Addison-Wesley. (Chapter on problems of the movement of water and solutes.)

Sutcliffe, J. (1968) *Plants and water*. Edward Arnold. (Section on transpiration and movement of water through plants.)

Chapter 2

The cell and water

Summary of practical work

section *topic*
2.1   The effect, on isolated animal cells, of altering the
composition of the external medium

2.2   The effect, on plant cells, of altering the composition
of the external medium

2.3   The osmotic potential of plant cells

2.4   The significance of turgid cells in the non-woody
plant

If an organism is to survive, there must be a balance between
the water taken in and given out. Water passing through the
xylem tissues of a plant or the gut of an animal is only the
first stage of the process. What happens in the individual
cells?

In many marine animals this is really no problem; the concentration of solutes in the cells is the same as that in seawater. Some animals, such as jellyfish, possess a transport
system which uses seawater as a fluid. In the open sea the
changes of solute concentration, and consequently water
concentration, are minute. However this is not so nearer the
coasts. Where a river enters the sea, there will be a region of
brackish water. Animals living in these waters, such as the
ragworm *Nereis diversicolor* and the crab *Carcinus maenas*,
are exposed to wide changes of water concentration in their
external medium. This may be almost fresh water at low
tide and normal sea water at high tide.

Such animals must either possess cells which can adapt to differences of concentration or they must be able in some way to regulate their water concentration.

Only animals which can survive these conditions can colonize an estuary and this is the main factor giving rise to the characteristic fauna of such areas.

Figure 10
Diagram showing the distribution of three species of *Gammarus* in an estuary.
After Serventy, Internationale Revue der gesamten Hydrobiologie und Hydrographie, **32**, *p. 286, Akademie Verlag, as modified in Yonge, C. M. (1949) New Naturalist Library*, The sea shore, *Collins*.

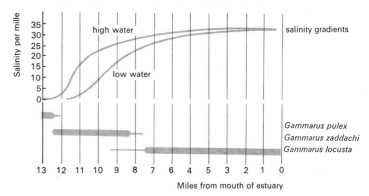

Figure 10 shows the distribution of three species of *Gammarus* in the estuary of the River Deben, Suffolk. Salinity is the major factor controlling their distribution.

Figure 11
Simplified diagram of an animal cell.

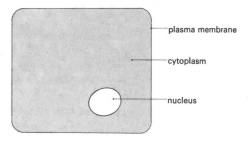

Water relations of an animal cell

Figure 11 is a simplified diagram of an animal cell. The cytoplasm is surrounded by a plasma membrane which exhibits special properties. One of these is that certain molecules can easily pass through it, whereas the passage of others is restricted or entirely prevented. Such a membrane is said to show differential permeability, and membranes of this type are to be found in all living cells. A simple view of these membranes is to regard them as sieves, allowing small molecules to pass through, but restricting or preventing the passage of larger ones. (Figure 12.)

Figure 12
Diagram of a hypothetical molecular sieve.

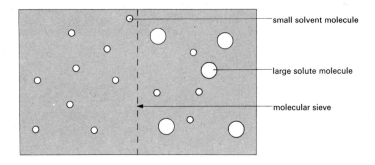

Although evidence exists to show that plasma membranes are probably more complex than this, the work in this chapter will be based on the assumption that the membranes function as simple sieves.

Diffusion and the cell

Molecules in a fluid system exhibit diffusion. This is due to the random movement of molecules in the system. The net result of this random movement is that molecules of a particular substance become distributed so that their concentration is uniform throughout the system.

Figure 13
'Model' cell in distilled water showing different concentrations of 'water' and solute molecules.

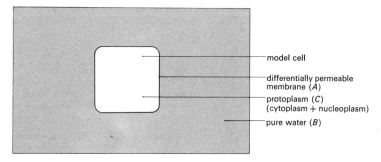

Suppose we examine the model illustrated in figure 13. Here we have a model animal cell with a differentially permeable membrane (A) enclosing protoplasm (C) in an external medium composed of pure water (B).

1   The concentration of water outside the cell will be 100 per cent.

2   Inside the cell the protoplasm consists largely of water, with various organic and inorganic molecules dissolved in it. The concentration of water here must therefore be less than 100 per cent.

3   The differentially permeable membrane will only permit the passage of small molecules such as water.

The water molecules in (B) will diffuse through the membrane, creating a diffusion pressure d_C, and so will the water in (C), in the opposite direction, giving a diffusion pressure d_B. But because the concentration of water is lower in (C) than in (B), there will be a difference or deficit.

$$d_B - d_C = \text{DPD } (\textit{diffusion pressure deficit})$$

Thus, although diffusion will occur in both directions, there is a net diffusion of water (solvent) molecules from (B) to (C). This is called osmosis. As (B) is pure water, the pressure required to stop osmosis is termed the *osmotic potential* of the solution inside the cell. We could say 'osmotic pressure' but, when comparing solutions with regard to their osmotic properties, 'pressure' could be a misleading word.

The net movement of solvent molecules will continue until either there is no diffusion pressure deficit, i.e. $d_B - d_C = 0$, or the pressure inside the cell prevents the inward diffusion of more water molecules.

2.1 The effect, on isolated animal cells, of altering the composition of the external medium

We can test our theoretical model of the osmotic relationships of an animal cell experimentally, if we expose isolated animal cells to different concentrations of water in their external environment. Red blood cells are convenient for this purpose.

Procedure

1　Prepare a stock solution by diluting citrated blood 1:10 with isosmotic saline. Place a 1 cm³ portion of this in each of 3 clean dry test-tubes labelled (a), (b), and (c).

2　Pour in 10 cm³ of each of the following reagents into each of the labelled tubes: distilled water; isosmotic saline; molar saline.

3　After a few minutes compare the turbidity of the solutions. You can do this either by using a colorimeter or by placing a sheet of small typescript behind the tubes and observing how easily it can be read. Use a suitable standard for comparison, such as a similar volume of the diluted blood.

4　Place a drop of liquid from the first tube on a microscope slide, place a clean cover-slip on top, and examine it under the microscope.
Repeat this with the other two tubes.

Questions

a   What was the effect of each of the reagents on the red blood cells?
b   How can the observed effects be explained in terms of the hypothetical model discussed?
c   What does this investigation suggest about the composition of the natural external medium of red blood cells?

Figure 14
Simplified diagram of a plant cell.

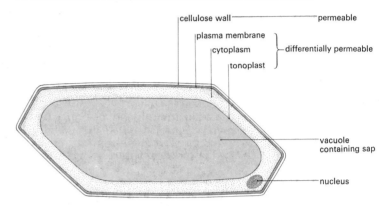

Plant cells and osmosis

The plant cell differs from an animal cell in that almost all plant cells possess an elastic and tough wall constructed of cellulose.

The cellulose cell wall is freely permeable to water and solutes, but is strong enough to resist high hydrostatic pressure.

Suppose such a cell is immersed in distilled water. There will be, as before, a higher water concentration outside the cell than inside and thus, as before: $d_B - d_C = DPD$.
Here too there will be a net diffusion of water into the cell. However, as the volume of the vacuole and cytoplasm increases the cellulose wall will be stretched. The wall will exert a force, called wall pressure, resisting stretching. The situation is analogous to a football being pumped up. The outer cover stretches slightly but then resists further stretching. More air being pumped in only raises the air pressure inside the bladder. In our model cell the situation is similar. Hydrostatic pressure, or *turgor pressure*, builds up inside the cell. Providing the wall is sufficiently strong, the hydrostatic pressure will become large enough to prevent the entry of more water by osmosis.

Figure 15
Osmotic potential and the cell.
Based on Baron, W. M. M. (1967)
Organisation in plants,
Edward Arnold.

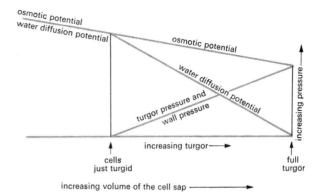

The changes in our model cell are illustrated graphically in figure 15.

When the cell wall is stretched to its greatest extent, wall pressure is equal and opposite to turgor pressure. Such a cell is said to be turgid.

If the cell is not completely turgid, then more water can be taken up. The amount which can be taken into the cell will depend on the difference between the osmotic potential and wall pressure. This quantity is known as the *water diffusion potential* or *water potential*.
Osmotic potential − wall pressure = water potential.

2.2 The effect, on plant cells, of altering the composition of the external medium

This time we can test our hypothetical model by exposing suitable plant cells to differing concentrations of water in their external environment. Isolated plant cells are not as convenient for use as a tissue composed of a single layer of cells. Such tissues are found as the outer epidermal layer to many plant organs.

Procedure

1   For this investigation it is essential to use a piece of plant tissue only one cell thick. The inner epidermis of a fleshy scale leaf of an onion bulb or a strip of coloured epidermis of rhubarb petiole are two suitable tissues.

2   Strip the tissue off the plant material and check that it is largely composed of a single cell layer. Cut it into small pieces about 5 mm^2 and place one piece each in watchglasses (a) and (b). One will contain distilled water and the other, molar sucrose solution.

3   Leave for 5–10 minutes, and then remove the pieces and make a temporary preparation for microscopical examination, mounting the tissue in the same liquid. Examine the cells under the microscope.

Questions

a In what ways do the plant cells respond differently from the animal cells?

b Is there any evidence of permeability in the cellulose cell wall to water and solute?

c How can you explain your observations in terms of the hypothetical model illustrated in figure 15?

2.3 The osmotic potential of plant cells

The osmotic potential of plant cells will obviously affect the water relations of the plant by altering the amount of water which can be taken in from the environment of the cell. The osmotic potential depends on the total number of solute molecules present, but not on their nature. This is an extremely difficult determination to make on a normal plant cell but it is possible to use indirect methods.

Procedure

1 Prepare a series of sucrose solutions of different concentrations 0·2 M to 1·0 M. The most convenient way is to put distilled water in one burette and molar sucrose solution (342g/dm³) in another and make up 0·2 M, 0·4 M, 0·6 M, and 0·8 M solutions by running appropriate quantities into specimen tubes.

2 Prepare 5 mm² pieces of epidermal tissue one cell thick as in 2.2. Place 3 pieces in each solution. Swirl the tubes to ensure that the epidermal pieces are completely immersed. Cork the tubes and leave for 20 minutes.

Figure 16
Epidermal cells of rhubarb.
(× 140)

 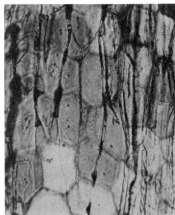

a Plasmolysed.

b Unplasmolysed.
Photo, Brian Bracegirdle.

3   After this interval, remove the pieces from the solutions one at a time. Mount each piece in a few drops of the solution in which it was immersed. Place a cover-slip on top and examine the tissue under the high power objective. Record the total number of cells visible in the field and also those which are plasmolysed. These are cells where the cytoplasm has come away from the cell wall, as shown in figure 16.
Move the slide to view another part of the piece and repeat. Continue this until you have counted cells from 5 or 6 different parts of the epidermis. Express your results as a percentage of plasmolysed cells.

4   Record the temperature of the solutions used (i.e. room temperature).

5   Plot a graph of percentage of cells plasmolysed against molar concentration of sucrose.

Questions

a   When 50 per cent of the cells in a piece of tissue are plasmolysed, what is the average wall or turgor pressure within the cells?

b   At 50 per cent plasmolysis what is the relationship between the osmotic potential of the contents of the cell and the diffusion pressure deficit?

c   What strength of sucrose solution shows an osmotic potential equivalent to that of the average potential of the cell vacuoles?

d   In this investigation we are using pieces of isolated tissue. Why might this give an incorrect finding for the cells *in situ* in the plant body?

e   A molar sucrose solution at 0° C (273° absolute) produces an osmotic potential of 22·4 atmospheres.
At room temperature the osmotic potential is given by:

$$\frac{(273 + \text{room temperature in } °\text{C})}{273} \times \text{osmotic potential at } 0°\text{C}.$$

Calculate the osmotic potential in atmospheres of the cells in the plant tissue that you used.

2.4 The significance of turgid cells in the non-woody plant

We have seen (Chapter 1) that non-woody plants wilt when there is a shortage of water. Our hypothesis is that, when water is taken up by plant cells or lost from them, there will be volume changes in the cell (figure 15). If these changes occur in individual cells, will they affect the dimensions of plant tissue as a whole? We can investigate this by using homogeneous non-woody plant tissue such as potato tubers or, alternatively, storage roots.

Procedure

1   Prepare a series of 9 containers with a range of sucrose solutions (0·1 M to 1·0 M) in the way described in 2.3. The liquid should be deep enough to immerse the potato cylinders completely.

2   With a cork borer about 10 mm in diameter, cut cylinders as long as possible from a large potato. Push the cork borer straight through the tuber; do not twist it or this will break the cylinder. Push the contents of the borer out carefully with a wooden dowel or the uncut end of a pencil.

3   Cut the ends of the cylinders so that their surfaces are at right angles. It is easiest to keep to one standard length of cylinder. This should be as long as possible and, in any event, not less than 5 cm.

4   Measure the length of each cylinder, using callipers or dividers; if you use the latter, bend the points inwards. Put 4 or 5 of the potato cylinders in each container. Make sure that you can identify the individual cylinders.

5   After a suitable time, not less than one hour, remeasure the length of each cylinder.

6   Express your results graphically, showing the percentage of change in length against the molar strength of the solution.

Questions

a   Is the hypothesis confirmed that when the content of water of individual cells is changed, this affects the dimensions of the plant tissue as a whole? Can you give a theoretical explanation for your finding?

b   From your results, does there appear to be any clear relationship between changes in cylinder length and the concentration of solutions bathing the cylinders?

c   Can you suggest anything occurring in the life cycle of a normal plant, which corresponds to the changes in length of potato cylinders, induced experimentally?

d   Instead of measuring the lengths of potato cylinders, what other simple measurement could you make which would indicate changes in their water content?

e   Using suitable pieces of homogeneous non-woody plant tissue as above, can you devise an experiment to see if there is a relationship between the strength of a piece of tissue and its water content?

Bibliography

Baron, W.M.M. (1967) *Water and plant life*. Heinemann Educational Books. (Section on the mechanism of water uptake.)

Steward, F.C. (1964) *Plants at work*. Addison-Wesley. (Section on plants in relation to water and solutes.)

Sutcliffe, J. (1968) *Plants and water*. Edward Arnold. (Section on vacuolated plant cell.)

Synopsis

1 Mammals are capable of maintaining an internal water balance under diverse environmental conditions.

2 The kidney has been established as the most important organ for maintaining water balance control.

3 The blood supply to the kidney is suited to the organ's functioning.

4 The nephron is the unit of function in the kidney.

5 Regions of the nephron associated with particular functions also possess a characteristic histological structure.

Chapter 3

Control by the organism

Summary of practical work

section   topic

3.1   Relationship of the urinary system to other organs of the body

3.2   Injection of the arterial blood system in the kidney

3.3   The histological structure of the nephron

The water balance in mammals

Mammals as a group are almost entirely terrestrial animals. They are very widely dispersed and are found on land over practically the whole globe. The only land mass where there are no indigenous mammals is Antarctica, although even here, an introduced species, man, can survive. Because the environmental conditions of the various habitats differ widely, the problems of water balance will also vary widely.

For example, the natural habitat of the Mongolian gerbil is the Gobi desert, where the rainfall seldom exceeds 20 cm per year and the daily temperature range can exceed 50°C. We can compare these conditions with those that surround a mammal such as the sloth, living in a tropical rain forest where the annual rainfall may exceed 2 m per year and the extreme temperature range may be only 10°C.

Irrespective of habitat the sources of gain and loss of water in the body are very similar. Table 1 shows what these are for man.

Table 1
Water balance. The figures given are in dm³ and are representative values of an adult for a 24 hour period in a temperate climate.
Adapted from Lippold, O. C. J. and Winton, F. R. (1968) Human physiology, *6th edition,* Churchill.

Gains			Losses	
Food			Urine	1·5
a	Water contained in food	1·0	Faeces	0·1
b	Water produced during metabolism of food in body	1·4	From lungs by evaporation	0·4
	Drink	1·5	From skin by sweating	0·9
		──		──
		2·9		2·9

The control which mammals can exert on the sources of loss of water is by no means perfect. A mammal cannot reduce the water loss via ventilation, as the internal surface of the lungs must be kept moist in order that gaseous exchange can effectively take place. Therefore, unless the relative humidity of the air drawn into the lungs is 100 per cent, a certain amount of moisture must be passed out with the exhaled air. Moreover, many mammals do not possess a complex sweat gland system such as is found in man. Where there are sweat glands they function automatically to control body temperature. This is accomplished by evaporation of water and consequently reduces the quantity of water in the body.

The most important organ for maintaining water balance within the tissues is the kidney.

Internal environment

More than a century ago the great French physiologist Claude Bernard developed the concept of the internal liquid environment, the 'milieu intérieur', of an animal. This consists of the blood, lymph, and other body fluids.

Bernard wrote: 'In experimentation on organic bodies, we need take account of only one environment, the external cosmic environment; while in the higher living animals, at least two environments must be considered, the external or extra-organic and the internal or intra-organic environment. . . .

'Indeed the internal environment of living beings is always in direct relation with the normal or pathological manifestations of organic units. In proportion as we ascend the scale of living beings, the organism grows more complex; the organic units become more delicate and require a more perfected internal environment. The circulating liquids, the blood serum, and the intra-organic fluids all constitute the internal environment.'

Figure 17
Dissection of a rat, with the alimentary canal exposed.

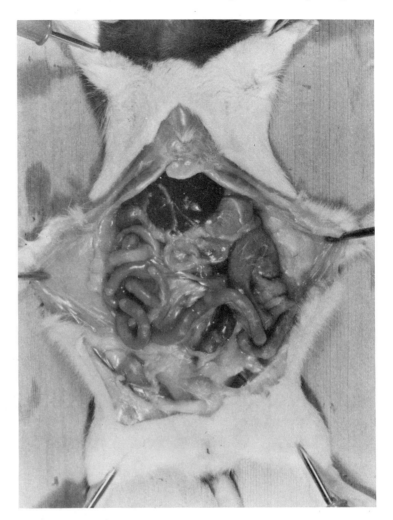

Bernard's hypothesis was that various organs of the body contributed to maintaining the constancy of composition of the liquids in this system. This hypothesis has since been found to be correct. The fact has now been established that the kidney is an organ with a major role in maintaining a stable internal environment, not only for water but for many other materials. Many of the kidney's functions are clearly illustrated in an analogue, the artificial kidney. (See the Topic Review, *The artificial kidney.*)

3.1 Relationship of the urinary system to other organs of the body

The kidneys and their ducts form the urinary system of the body. The structural relationship of this system to the rest of the body is very similar in mammals, and therefore any convenient mammal such as a rat or mouse is suitable for this investigation.

Procedure

1   Pin the dead mammal down, through the limbs with the ventral side uppermost. Carefully open the animal with a ventral abdominal incision, cutting through the skin and body wall, to expose the alimentary canal as shown in figure 17.

2   Remove the alimentary canal by cutting through the oesophagus and rectum, but leaving about 2 to 3 cm of rectum in position. Gently remove the rest of the canal from the body cavity by severing the mesenteries and blood vessels. Wash out the body cavity.

3   Locate the two kidneys, examine the blood supply to and from the kidney and trace the kidney ducts to the urinary bladder.

Figure 18
Dissection of a rat. The scissors are in position to cut one side of the pelvic girdle.

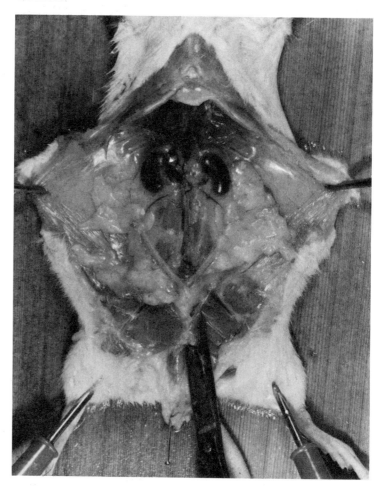

4   To follow the urinogenital system to the exterior, you need to remove part of the pelvic girdle. Locate the position of the pelvic girdle and then, holding a pair of scissors with the blades nearly horizontal, cut through the pelvic girdle on one side, a little to the right of centre as indicated in figure 18. Then make a similar cut on the left side and remove the central cut portion. Pin out the knees so as to clearly expose the region and then trace the urethra to its external opening.

5   Make a quick labelled sketch or diagram to show the relationships of the various organs.

6   Carefully remove one kidney and the upper portion of its associated ureter. Slide the kidney from side to side, parallel to its dorsal surface so as to cut across the portion of the structure where the ureter leaves it. Wash the surface with water and examine with a hand lens. (Figure 19.)

Figure 19
The structure of the human kidney.

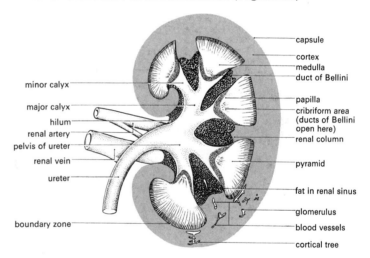

Questions

a   What is the relationship between the kidneys and the other organs in the body cavity?

b   From what part of the kidney does the ureter leave and how and where does this join onto the urinary bladder?

c   From where does the artery supplying the kidney come and to where does the vein from the kidney go?

d   From the relationships of the arterial blood supply what hypothesis can you make about the pressure of the blood entering the kidney?

Kidney functioning

The kidney maintains the water balance of the body by altering the amount and composition of the urine produced. In order to appreciate the way in which the kidney produces urine it is important to understand the micro-anatomy of the organ. We can first investigate the course of the arterial blood in the kidney by means of an injection technique.

3.2 Injection of the arterial blood system in the kidney

Kidneys can be injected with colouring matter through the main blood vessels to provide a more or less permanent record of the internal blood supply. The simplest procedure is to inject coloured latex into the main artery.

Procedure

1   For this investigation you require kidneys fresh from a butcher. It is best to obtain kidneys still enclosed in fat. Remove the fat and examine the kidney to locate the renal artery.

2   Inject into this 3 to 5 cm^3 of warm Ringer's solution. A convenient way of doing this is to use a 5 cm^3 hypodermic syringe, fitted with a short length of thin rubber tubing, and with a short piece of glass tubing at the end, small enough to fit the artery.

3   Using a similar process follow the first injection into the artery with a similar quantity of warm red latex. Inject this slowly at the rate of about 2 cm^3 per half minute. Then place the kidney in 2-normal hydrochloric acid for 24 hours.

4   Slice the kidney in the flat plane, parallel with the exit of the ureter. Examine it under a low power stereoscopic binocular microscope.

5   Select a portion of the cortex showing the injected blood vessels clearly. With a scalpel slice a section about 3 mm thick parallel to the flat surface.

6   Place this section in fresh 2 per cent pepsin in a specimen tube. The hydrochloric acid already present should be sufficient to bring the pH down to about 2 to 3. Close the tube and leave for 2 to 3 days. After this time remove the section and rinse it gently in water. Using a pair of forceps pick out suitable blood vessels and mount them in water on a slide. Examine under a microscope.

Questions

a   What parts of the kidney structure can you identify? Make a sketch of these.

b   Compare the distribution of the branches of the renal artery supplying the glomerulus with a conventional diagram of the structures. In what way does the distribution in your preparation differ from the diagram?

c   What can be inferred from the preparation about the resistance to flow in the glomerulus blood vessels?

d   Does the preparation give any indication of the nature of the barrier between the blood vessels of the glomerulus and the lumen of Bowman's capsule?

The nephron

The unit of kidney function is the kidney tubule or nephron. Figure 20 shows its general structure.

Each nephron consists of
renal (Malpighian) corpuscle
renal tubule

Figure 20
Dissected nephrons.

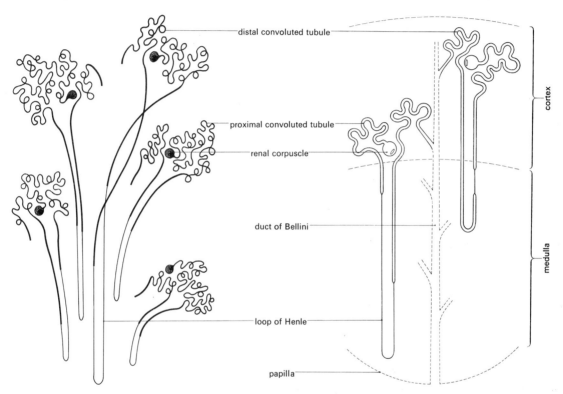

The length of a nephron is between 2 and 4 cm. As there are about one million of them in each kidney of man the total length of tubule in the two kidneys has been calculated to be between 50 and 100 miles.

Structure of the nephron

In the middle of the seventeenth century a young Italian, Marcello Malpighi, first discovered the tubular nature of the kidney. This linked with a discovery a few years before by a young student, Lorenzo Bellini, of the hollow ducts found in the cut surface of a kidney. These ducts are named after him.

Malpighi further discovered the tufts of blood capillaries which he called corpuscles. It was not until 1842 that William Bowman, a demonstrator at King's College, London, completed the picture by discovering the relationship between the Malpighian corpuscle and the kidney tubule via Bowman's capsule.

Figure 21
A renal corpuscle of a cat kidney.
a Photomicrograph of a silver preparation. (×220)
Photo, Brian Bracegirdle.

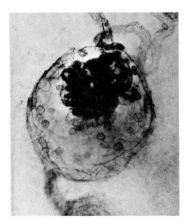

b Diagram.

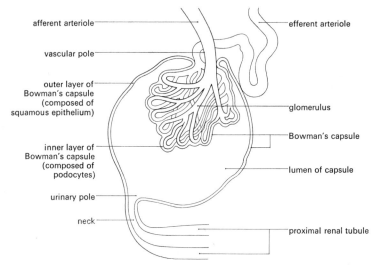

Figure 21 shows the structure of a renal corpuscle.

Functioning of the renal corpuscle

For almost a hundred years a controversy continued about the functioning of the renal corpuscle. In the early 1920s A.N. Richards and his co-workers used a new technique, employing the micropipette, to sample the liquid in the Bowman's capsule of a functioning frog kidney. They chose a frog because this animal possesses a large capsule.

Figure 22
Richardson's method of using a
micropipette to obtain a sample of
glomerular fluid from the Bowman's
capsule in a frog.
Based on Lippold, O. C. J. and Winton,
F. R., (1968) Human physiology,
6th edition, Churchill.

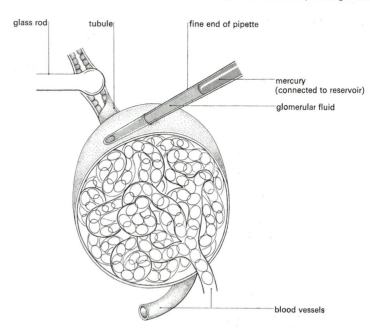

The glass rod is used to close the tubule. The end of the
micro-pipette (7 to 15μ in diameter) is inserted into Bowman's capsule. By slightly lowering a levelling bulb a
minute quantity of liquid could be drawn up into the pipette.
The investigators obtained a maximum of 1 mm³ of liquid by
this process and had to devise special methods of analysis.
Table 2 shows the analysis of this fluid compared with blood.

Table 2
The composition of plasma and
glomerular filtrate.

	g per 100 cm³	
	Plasma	Glomerular filtrate
Water	90–93	97–99
proteins	7–9	0
creatinine	0·001	0·001
glucose	0·1	0·98
urea	0·03	0·03
uric acid	0·002	0·002
ammonia	0·0001	0·0001
calcium	0·01	0·01
magnesium	0·002	0·002
potassium	0·02	0·02
sodium	0·3	0·31
chloride	0·35	0·37
phosphate	0·003	0·003
sulphate	0·003	0·003

From the nature of the liquid found in Bowman's capsule, what, do you suggest, is the mechanism that was operating as the blood circulated through the glomerulus? Later, the investigators obtained similar results with mammalian kidneys.

Although these results indicated the process by which the liquid was formed in Bowman's capsule they did not show that sufficient pressure was available for the process to take place. The nature of the arterial blood supply appears such that blood would reach the capillary network forming the glomerulus at high pressure. Is this a sufficient force to drive the process?

In the renal corpuscles three pressures will be operating, as shown in table 3.

Table 3
Forces involved in glomerular filtration.

	Force in mm of mercury
mean arterial pressure	100
pressure in glomerular capillaries	70
pressure in Bowman's capsule	15
therefore net filtration pressure	$(70-15)=55$
osmotic potential in blood due to colloids	30
therefore net pressure available to overcome frictional resistance of membrane	$55-30=25$

Functioning of the tubule

It has been estimated that the amount of water passing through walls of Bowman's capsules in the kidneys of man per 24 hours averages 170 dm^3. The amount of urine produced per day is almost 1500 cm^3. Thus, somewhere in the system, more than 99 per cent of the water must be reabsorbed.

Proximal tubule

By means of chemical analysis and investigations with the micro-pipette, it has been established that water and dissolved solutes are absorbed at osmotically equal rates in the proximal tubule. Thus the tubule neither concentrates nor dilutes the liquid inside it. A large proportion of the liquid passing down the proximal tubule is absorbed. Figure 23 illustrates the micro-anatomy of the proximal tubule.

Figure 23
Proximal tubule cells.

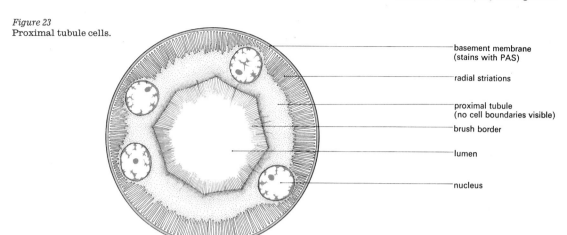

basement membrane
(stains with PAS)

radial striations

proximal tubule
(no cell boundaries visible)

brush border

lumen

nucleus

a   Transverse section of a proximal tubule as seen by light microscopy

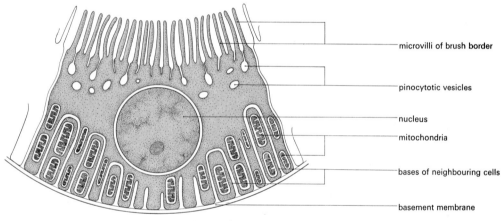

microvilli of brush border

pinocytotic vesicles

nucleus

mitochondria

bases of neighbouring cells

basement membrane

b   A proximal tubule cell as seen by electron microscopy

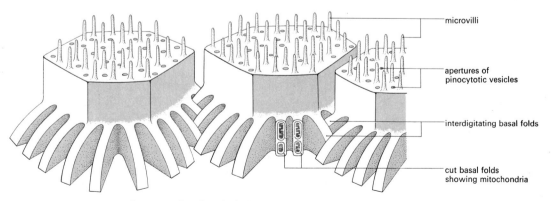

microvilli

apertures of
pinocytotic vesicles

interdigitating basal folds

cut basal folds
showing mitochondria

c   A reconstruction of proximal tubule cells

What evidence do these diagrams provide that the proximal tubule might be concerned with the passive absorption of materials? Is there any evidence that other processes might be taking place?

Loop of Henle and distal tubule

Analysis has shown that the liquid entering the loop of Henle has the same concentration as that leaving Bowman's capsule. It was thought for a long time that the loop of Henle was an important structure for reabsorbing water.

1 Only birds and mammals possess loops of Henle. Both groups are the only ones capable of producing a urine more concentrated than the body fluids (hyperosmotic).

2 The length of the loop can be correlated with the habitat of the animal and the urine concentration finally produced. For example: The beaver has only short loops of Henle; its habitat is aquatic (600 milliosmoles).

The rabbit has long and short loops of Henle (1500 milliosmoles).

The desert rat has long loops of Henle; its habitat is very dry (6000 milliosmoles).

(An osmole = 1 mole of solute per dm³.)

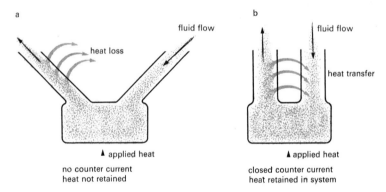

Figure 24
Heating units.
a With the inlet and outlet pipes apart.
b By control, a counter current flow system.
After Lockwood, A. P. M. (1963) Animal body fluids and their regulation, *Heinemann.*

Unfortunately, micropuncture techniques shattered this theory. It was found that the fluid leaving the loop of Henle could be *less* concentrated than that entering it.

Counter current flow

Figure 24 shows the outline of a counter current flow system used in engineering to produce the most effective use of a heat source. In (a), the inlet and outlet pipes are apart so that any heat lost from the outlet pipe is lost completely. As a contrast, in (b), much of the heat lost from the outlet pipe is transferred to the inlet pipe to preheat the water. The loop of Henle provides a similar type of structure although in this case the counter current exchange is of concentration, not temperature. The arrangement in the loop of Henle is shown in figure 25.

Figure 25
Diagram to show the changes in
osmotic concentration of tubular
fluid as it passes through the loop of
Henle and the rest of the nephron.

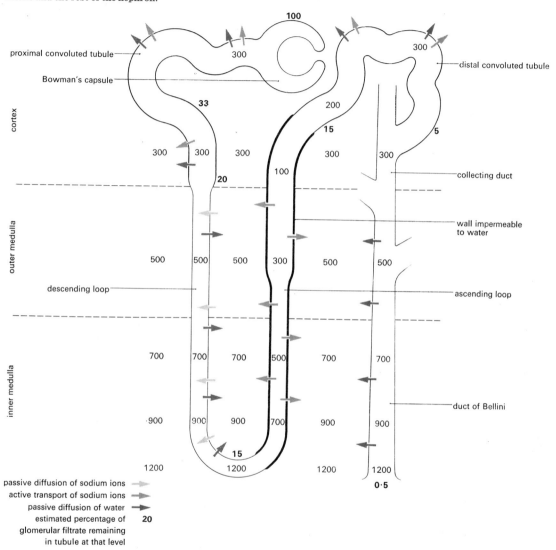

Questions

Examine figure 25.

a   What is the concentration of tubular fluid at the base of the
    loop of Henle?
b   What effect on the composition of the tubular fluid is created
    by the region of the loop impermeable to water?
c   What is the proportion of water removed from the filtrate
    during its passage from Bowman's capsule to the pelvis of
    the kidney?

d   Where is the urine finally concentrated?
e   What process do you think causes this concentration?
f   What advantage is gained by maintaining such a high osmotic concentration in the tissues surrounding the base of the loop?
g   The tissues around the loop of Henle must be supplied with blood. In order to maintain the high osmotic concentration in them, how do you think the blood vessels might be arranged?

3.3 The histological structure of the nephron

As various parts of the nephron perform different functions we might expect the tubule to show a different structure along its length. Unfortunately as the nephron is a three dimensional structure, it is impossible to see all of the structure from the examination of one section. To do this fully you would need to use a series of sections cut consecutively through the kidney and build up a reconstruction of a nephron by a lengthy study of these.

Figure 26
Diagram of a nephron.
Based on Freeman, W. H. and Bracegirdle, B. (1966) An atlas of histology, Heinemann.

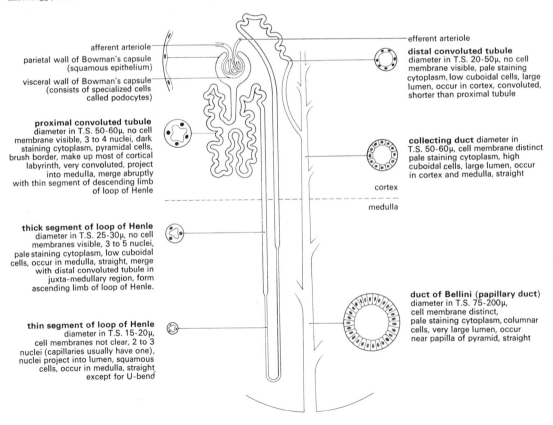

afferent arteriole

parietal wall of Bowman's capsule (squamous epithelium)

visceral wall of Bowman's capsule (consists of specialized cells called podocytes)

proximal convoluted tubule
diameter in T.S. 50-60µ, no cell membrane visible, 3 to 4 nuclei, dark staining cytoplasm, pyramidal cells, brush border, make up most of cortical labyrinth, very convoluted, project into medulla, merge abruptly with thin segment of descending limb of loop of Henle

thick segment of loop of Henle
diameter in T.S. 25-30µ, no cell membranes visible, 3 to 5 nuclei, pale staining cytoplasm, low cuboidal cells, occur in medulla, straight, merge with distal convoluted tubule in juxta-medullary region, form ascending limb of loop of Henle.

thin segment of loop of Henle
diameter in T.S. 15-20µ, cell membranes not clear, 2 to 3 nuclei (capillaries usually have one), nuclei project into lumen, squamous cells, occur in medulla, straight except for U-bend

efferent arteriole

distal convoluted tubule
diameter in T.S. 20-50µ, no cell membrane visible, pale staining cytoplasm, low cuboidal cells, large lumen, occur in cortex, convoluted, shorter than proximal tubule

collecting duct diameter in T.S. 50-60µ, cell membrane distinct pale staining cytoplasm, high cuboidal cells, large lumen, occur in cortex and medulla, straight

cortex

medulla

duct of Bellini (papillary duct)
diameter in T.S. 75-200µ, cell membrane distinct, pale staining cytoplasm, columnar cells, very large lumen, occur near papilla of pyramid, straight

Procedure

1   Figure 26 is a diagrammatic reconstruction of a nephron providing data on the various types of tubule cell. Use this with your slide as an aid to identification.

2   Examine a microscope slide preparation of kidney. First search the slide under low power, identify the cortex and medulla and any notable structures such as large blood vessels.

3   Then carefully examine regions of the cortex and medulla under high power. Try to locate sections of the different parts of the tubule.

4   Having identified a structure, make a diagram of it and tabulate the appropriate data alongside this.

Questions

a   By examining sections of kidney, what can you deduce about the function of various parts of a nephron?

b   How could you obtain more precise evidence about the function of various parts of the tubule?

c   We have said that the kidney is the most important organ for maintaining an internal water balance in the body. As fluid passes down the nephron about 99 per cent of the water is reabsorbed. Why is this not sufficient to explain how the kidney regulates the water balance? The composition of normal urine is given in Table 4.

Table 4
Typical components of normal urine in man.

	g per 100 cm^3
Water	96–97
creatinine	0·15
urea	2·0
uric acid	0·05
ammonia	0·05
calcium	0·015
magnesium	0·01
potassium	0·15
sodium	0·35
chloride	0·6
phosphate	0·12
sulphate	0·18

Apart from water reabsorption, what other processes must be taking place during the passage of fluid along the nephron?

Bibliography

Bernard, C. (1865) trans. Copley Greene, H. (1957) *An introduction to the study of experimental medicine.* Dover (Part 2, Chapter 3.)

Best, C.H. and Taylor, N.B. (1958) *The living body.* Chapman & Hall. (Chapter on the physiology of the kidney.)

Clegg, A.G. and Clegg, P.C. (1963) *Biology of the mammal.* Heinemann Medical Books. (General account of kidney function.)

Lockwood, A.P.M. (1963) *Animal body fluids and their regulation.* Heinemann Educational Books. (General account of body fluid regulation with chapter on kidney functioning.)

Nuffield Advanced Biological Science (1970) Laboratory Guide, *Maintenance of the organism.* Penguin. (Chapter 2, 'Gas exchange systems.')

Sands, M.K. (1970) Nuffield Advanced Biological Science Topic Review *The artificial kidney.* Penguin.

Smith, H.W. (1953) 'The kidney'. *Scientific American* Offprint No. 37. (Interesting account of discovery of kidney structure and functioning, together with evolutionary ideas.)

Synopsis

1 Water uptake may be influenced by the metabolism of living cells.

2 The presence of growth substances may affect water uptake.

3 The effect of temperature change on rate of water uptake provides evidence to establish the nature of the process.

4 By use of radioactive isotopes and autoradiography the uptake of salts can be followed in plant tissue.

Extension work I

Water and salt uptake by plants

Summary of practical work

section   *topic*

E1.1   The effect of a growth substance

E1.2   Colorimetric measurement of the uptake of water

E1.3   The use of phosphorus 32

Water uptake in plants

In Chapter 1 we investigated the role of the root, stem, and leaf in the water relations of a plant and also the effect of environmental conditions around the aerial shoot. Many of the phenomena connected with the water relations of plants can be explained in terms of purely physical processes, but in this chapter we examine evidence that the processes involved may be more complex than this.

E1.1 The effect of a growth substance

A large number of compounds are said to promote or induce growth in plant organs. The way in which they work is uncertain but it may be connected with water uptake. We can study this by repeating investigation 2.4, 'The significance of turgid cells in the non-woody plant', but keeping the cylinders of plant tissue in solutions containing a growth substance.

Procedure

Carry out the procedure given under 2.4 but with the follow-ing alteration.

Prepare the range of sucrose solutions by making up 25 cm³ at double strength (0·2 M to 2·0 M) and then adding to each 25 cm³ of a solution containing 10 parts per million of indole acetic acid (IAA). In this way you will have solutions of the same strength as before but each containing 5 p.p.m. of IAA.

Questions

a   Express your results graphically and compare them with those obtained without the addition of IAA. Does IAA appear to affect the changes in cylinder length brought about by the different sucrose solutions?

b   If there are changes due to the presence of IAA, are the effects of osmosis increased or decreased by its presence?

c   Suggest hypotheses to account for your findings and a means by which at least one of them could be tested.

The effects of temperature

Our knowledge of the effects of temperature on enzymes might provide us with useful evidence for or against the idea that the uptake of water by plant tissues is a purely physical process. Boiling is one of the tests used for the presence of enzymes, since it inactivates them. However, it also destroys cell membranes, so it could not provide useful information about water uptake.

There appears to be a general rule that the Q_{10} of physical processes is about 1, whereas the Q_{10} of chemical (and bio-chemical) processes is nearer 2.

$$Q_{10} = \frac{\text{rate of reaction at } (t° + 10° \text{C})}{\text{rate of reaction at } t° \text{C}}$$

Some typical examples are given in table 1.

Table 1

	Reaction	Q_{10}
Physical	solubility of carbon dioxide in water	0·72
	diffusion of potassium chloride in agar gel	1·22
Chemical	hydrolysis of starch	1·96
	hydrolysis of maltose by maltase	1·90
	hydrolysis of peptone by erepsin	2·60

The measurement of water uptake at two temperatures, 10° C apart, might therefore give us useful evidence.

E1.2 Colorimetric measurement of the uptake of water

If cell membranes are differentially permeable, water will pass into the cell, leaving larger molecules outside. If these are coloured, then the density of colour should increase as water is taken up. We could then measure the increase by a colorimeter.

With most coloured salts this idea will not work because, in solution, they have a considerable osmotic potential and their presence outside living tissue will oppose the entry of water by osmosis.

The osmotic potential of a solution depends on the number of solute particles present and not on their nature.

There are, however, soluble coloured substances which have exceedingly large molecules (for example, blue dextran has a molecular weight of about 2000000). Dilute solutions of these substances have very low osmotic potentials and do not oppose the entry of water into tissues.

Procedure

1   Cut out cylinders of potato with a cork borer (8–10 mm in diameter) and chop them into pieces exactly the same length (about 5 mm). This can be done by holding two razor blades, separated by a piece of wood 5mm wide. You require 100 pieces for this experiment.

2   Put all these cylinders into 0·4 M sucrose solution for 1 hour.

3   Divide them into two groups of 50 each. Transfer these to two boiling tubes and wash away traces of sucrose solution by repeated rinsing with tap water.

4   Add exactly 25 cm³ of a solution of blue dextran 2000 containing 0·1 g per 100 cm³ of water. This quantity should just cover 50 potato cylinders in a boiling tube.

5   Withdraw blue dextran solution from one tube, transfer it to a colorimeter fitted with a red filter, and set the light to give full scale deflection. Check that the solution from the other tube gives the same reading. Cork both tubes.

6   Return the solution to immerse the potato tissue. Arrange one tube in a beaker of water at room temperature and the other in a water bath set 10° above room temperature.

7   Measure the optical density of each solution at 20 minute intervals.

Questions

a   Express your results graphically and determine the time taken for each solution to change from the initial optical density to that of the cooler solution at the end of the experimental period.

b   Calculate the Q_{10} from the rates of water uptake.

c   Does the Q_{10} obtained suggest that water uptake by potato tissue is essentially a physical process or a chemical one?

Chemical elements required by plants

The elements required in the largest amounts for plant growth–carbon, hydrogen, and oxygen–are absorbed as carbon dioxide, water, and oxygen gas. In the form of various compounds these make up the bulk of the weight of a plant.

Other elements are, however, essential for the healthy growth of plants. Nitrogen is needed for the production of amino acids and consequently for plant proteins. Five other elements are also required in relatively large quantities– calcium, magnesium, phosphorus, potassium, and sulphur. Plants also make use of other elements, known as minor nutrients, in small amounts down to a few parts per million. Amongst these elements are boron, chlorine, copper, iron, manganese, molybdenum, and zinc. Table 2 gives some data on the known uses of these elements.

These mineral nutrients normally enter the plant in the form of ions. As the quantities needed by a plant are small compared with the uptake of water, direct evidence by chemical analysis depends on very delicate and time consuming techniques. Today, the use of radioactive tracers has made this study a relatively simple process.

E1.3 The use of phosphorus 32

Phosphorus 32 is an isotope of phosphorus. Its radioactive property enables the element to be traced within an organism, for it behaves chemically in a fashion identical to normal phosphorus. Its half-life is 14·2 days, which means that half its radioactivity will have disappeared after that time. One quarter of the original radioactivity will be left after 28·4 days and so on. ^{32}P emits a beta (electron) radiation which is somewhat harder than carbon 14 which you may have used already.

The entry and distribution of radioisotopes into living tissues can be traced either with a Geiger-Müller (G.M.) tube or by means of autoradiography. The basic experimental technique is extremely simple, and the procedure that follows is one of many methods available.

Before you start, you must understand and conform to certain simple precautions necessary when dealing with any radioisotope source. The amount of ^{32}P used in this investigation does not constitute a hazard providing it does not enter the body via the mouth, an open wound, or any other means. Unlike poisons or explosives, radioactive chemicals cannot be 'neutralized' or made safe, and it is for this reason that the following rules must be obeyed.

Table 2
The minerals used by plants.
From Baron, W. M. M. (1967)
Organisation in Plants,
Edward Arnold.

Element	Form in which absorbed	Quantity utilized (approx.) as per cent of dry wt of plant	Functions	Effect of deficiency	Ecological and agricultural notes	Fertilizers
Major nutrients						
1 Nitrogen	NO_3^- (or NH_4^+)	3·5	Amino acids, proteins and nucleotides	Chlorosis, small-sized plants	Frequently deficient; organic manuring and addition of nitrogenous fertilizers often necessary	Ammonium sulphate, sodium nitrate, nitro-chalk (ammonium nitrate and fine chalk)
2 Potassium	K^+	3·4	Enzyme, amino acid and protein synthesis. Cell membranes. Increases vigour	Leaves have yellow edges; premature death	Plants need more potassium after heavy manuring with nitrogen and phosphorus. Most available on acid soils, though it may be leached out	Potassium sulphate
3 Calcium	Ca^{++}	0·7	Calcium pectate of cell walls. Development of stem and root apices	Stunting of the root and stem	Little present in acid soils. Has important effect on soil by assisting flocculation of clay particles	Nitro-chalk, calcium phosphate, basic slag, superphosphate of lime $[Ca(H_2PO_4)_2 + CaSO_4]$
4 Phosphorus	H_2PO_4 (ortho-phosphate)	0·4	Formation of 'high-energy phosphate' (ATP and ADP). Nucleic acids. Phosphorylation of sugars	Small-sized plants; leaves a dull, dark green	Frequently deficient. Little is available over pH 7	Superphosphate of lime, calcium phosphate, basic slag.
5 Magnesium	Mg^{++}	Small quantity	Part of the chlorophyll molecule. Activator of some of the enzymes in phosphate metabolism.	Chlorosis; older leaves turn yellow, their veins remain green	Often deficient on acid soils	Magnesium sulphate, basic slag
6 Sulphur	$SO_4^=$	Small quantity	Proteins which contain thiol (–SH) groups	Chlorosis	Seldom deficient in Great Britain due to sulphuric acid contained in atmospheric pollution	—
Minor nutrients						
7 Iron	Fe^{++}	Small quantity	Chlorophyll synthesis. Cytochromes	Chlorosis, young leaves turn yellow-white, their veins remain green	Much less available on calcareous soils, being in the form of insoluble ferric hydroxide	Chelated iron (in which the iron is bonded to an organic molecule), e.g. sequestrene 138 Fe

	Element	Form in which absorbed	Quantity utilized (approx.) as per cent of dry wt of plant	Functions	Effect of deficiency	Ecological and agricultural notes	Fertilizers
8	Chlorine	Cl^-	0·8	Osmosis and anion/cation balance; probably essential in photosynthesis in the reactions in which oxygen is produced	Effects slight	—	—
9	Copper	Cu^{++}	Trace (less than 0·0001 per cent)	Activator group of polyphenol and ascorbic oxidase enzyme systems	Shoots die back	Helps to improve soil condition	—
10	Manganese	Mn^{++}	Trace (less than 0·0001 per cent)	Activator of some enzymes (e.g. carboxylases)	Chlorosis and grey-specks on leaves	Manganese is in the divalent state in acid soils and is readily available even to a toxic level. At high pH manganese is in the trivalent state, which plants cannot make use of	—
11	Zinc	Zn^{++}	Trace (less than 0·0001 per cent)	Activator of some enzymes (e.g. carboxylases)	Leaf malformation	More often deficient on acid soils due to adsorption onto colloidal complexes in soil	—
12	Molybdenum	Mo^{+++} or $^{++++}$	Trace (less than 0·0001 per cent)	Nitrogen metabolism, enzyme nitrate reductase	Size of plants slightly reduced	If molybdenum is deficient, plants may survive if nitrogen is supplied as NH_4^+, but not if supplied as NO_3^-	—
13	Boron	$BO_3 \equiv$ or B^4O^7 (borate or tetraborate)	Trace (less than 0·0001 per cent)	Influences Ca^{++} uptake and utilization. Differentiation and pollen germination	Brown heart disease	Easily leached from soils, particularly those of low pH	—
14	Cobalt	Co^{++}	Trace (less than 0·0001 per cent)	Various roles in symbiotic nitrogen-fixing plants	—	—	—
15	Fluorine, nickel	F^- Ni^{++}	Trace (less than 0·0001 per cent	Not known, but possibly essential in some cases	—	—	—
	Non-essential elements						
16	Silicon	$H_2SiO_4^=$	1·0 (grasses)	Straw formation (calcium silicates). Not essential to most plants	Slight decrease in weight	—	—

	Element	Form in which absorbed	Quantity utilized (approx.) as per cent of dry wt of plant	Functions	Effect of deficiency	Ecological and agricultural notes	Fertilizers
17	Sodium	Na^+	Trace	Osmotic and anion/cation balance, probably not essential to most plants	Effects slight	—	—
18	Aluminium	Al^{+++}	Trace	Not essential. May cause upset to cell division system	—	More available in acid soils; it may prevent the growth of calcicoles	—

Laboratory rules to be followed when using radio-active compounds

1 Remove from the laboratory all unnecessary books and personal belongings which are not essential to the experiment to be performed.

2 Never eat or drink anything in the laboratory.

3 Always wear a laboratory coat (in the event of a spill, only one garment need then be destroyed).

4 Never use your mouth or tongue for pipettes, wash-bottles, or fixing labels.

5 When you are transferring radioactive material, do it over a plastic or metal tray.

6 Wear rubber or plastic gloves, but, even so, never undertake work with radioactive substances when you have any kind of wound below the wrist.

7 All gloves, apparatus, and instruments which may have been in contact with radioactive material must be carefully monitored with a proper survey instrument such as a Geiger-Muller tube and ratemeter. Such equipment must be placed in a properly labelled container at the end of the session.

8 Radioactive waste (such as plant tissue which has been exposed to ^{32}P and radioactive solutions) must be placed in a special container as directed by the teacher and *not* thrown into ordinary waste bins or sinks.

Procedure

1 Obtain two healthy plants of equivalent size and leaf number e.g. *Impatiens balsamina*, or tomato seedlings about 15 cm high, and carefully wash all traces of soil from their roots. Immerse the washed roots in water in a small container such as a 75 by 25 mm specimen tube.

2 You will probably use the ^{32}P in the form of a tablet. Pick up the packet containing the tablet with a pair of forceps and cut off a corner with a pair of scissors. Without touching the packet by hand, allow the tablet to fall into a beaker containing 20 cm³ of distilled water. Allow the tablet to dissolve. Collect some of the solution in a hypodermic syringe.

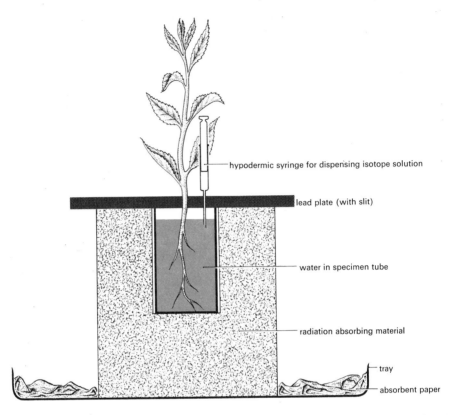

hypodermic syringe for dispensing isotope solution

lead plate (with slit)

water in specimen tube

radiation absorbing material

tray

absorbent paper

Figure 27
Use of ³²P.

Set up an apparatus of the kind shown in figure 27. The specimen tube (7 × 25 mm) should be of polythene.

3   Using the syringe, inject radioactive phosphate solution into the medium round the roots of both plants so that each vessel contains 0·25 micro-curies (μc) of activity.

4   Measure the background radiation above the lead cover (see figure 27), using a G.M. tube.

5   Cover one plant with a bell jar or plastic bag to diminish transpiration and provide a draught of air from a fan over the other plant. Provide artificial light equally over both.

6   Check radiation in the regions of both plants above the level of the lead cover after 5 to 6 hours. Take several readings and record the highest obtained in each case. Repeat this, if necessary, after a further 18 hours.

7   If the readings obtained are well above background level, you can prepare for autoradiography. To do this first cut a piece of cardboard to the same size and shape as the X-ray film or paper you will be using (see 9), e.g. approximately 12 by 16 cm. Remove the plant, cut off the roots, and arrange the shoot as you want it, on the sheet of cardboard. It may be necessary to cut the shoot and mount the pieces.

8 The shoot will not die immediately and there will be chemical reactions which might abnormally alter the distribution of the phosphate unless you stop them. To stop chemical reactions in the leaves, use a fixative as follows. Put large smears of polystyrene cement in appropriate positions on the card. Using forceps, place the shoot in position so that each leaf has a smear of cement under it. With a glass rod, gently press the leaves into the cement until it is evenly absorbed. The leaves will turn a darker shade of green.
Avoid touching the shoot by hand during these procedures. Allow the cement to dry and then insert the cardboard and attached shoot into a thin polythene or Cellophane bag of appropriate size. Seal it with tape and mark it *Radioactive*.

9 Take the sealed card to a darkroom and under suitable safe light, take it out of the bag and place it plant side down on the sheet of X-ray film or paper in a folder. Close the folder and take it to the fume cupboard or some other place specially reserved for radioactive material. Place a heavy weight such as one or two bricks on top of the folder to press film and card together, and leave it for a suitable length of time.

10 Take the folder to the darkroom and process the film according to the manufacturer's instructions. When you have answered the questions below, dispose of the card with the shoot (still in the bag) as directed.

11 Dispose of the remainder of the plant and the phosphate solution as directed. Monitor all apparatus and instruments which have been in contact with radioactive material.

Questions

a List the parts of the plants showing signs of phosphorus 32 as revealed by the autoradiographs.

b Do all parts of the leafy shoots appear to have received equal amounts of phosphorus 32?

c If not, can you suggest a hypothesis to account for the unequal distribution of the isotope?

d Are the autoradiographs of the two plants essentially the same or does the draught and still-air treatment appear to have affected the uptake of phosphorus-32?

e Do your experimental findings support the hypothesis that the uptake of salt depends on the stream of water passing through a plant?

f How could you modify this experiment in order to investigate the pathway of radiophosphorus from the roots into the leaves of a plant?

g Make a list of other conditions, besides the movement of surrounding air, which might have a bearing on the uptake phosphorus and which could be investigated by modifying the experimental procedure described.

Review

Question

Summarize the current ideas on the entry of water and salts into plants.

Bibliography

Baron, W.M.M. (1967) *Water and plant life*. Heinemann Educational Books. (Chapters on water relations and mineral nutrition.)

Nuffield Advanced Biological Science (1970) Laboratory Guide *Maintenance of the organism*. Penguin. (Chapter 7, 'Photosynthesis.')

Paice, P.A.M. (1968) 'Simple radioisotopic experiments in school biology.' *School Science Review* 49, 170, pp 62–78. (Details of techniques for radiobiology.)

Richardson, M. (1968) *Translocation in plants*. Edward Arnold. (Movement of minerals in plants.)

Steward, F.C. (1964) *Plants at Work*. Addison-Wesley. (Water relations and inorganic nutrition.)

Stoneman, C.F. (1970) Nuffield Advanced Biological Science Topic Review *Photosynthesis*. Penguin.

Sutcliffe, J. (1968) *Plants and water*. Edward Arnold. (Water relations of plants.)

Chapter 4

Stimuli and their influences

Summary of practical work

Stimuli

A change in the environment which produces an effect on an organism is said to be a stimulus, although all changes may not be due to stimuli from the external environment. Stimuli may affect organisms in different ways. They may initiate activity : a dog will run to the door when it hears the sound of known footsteps coming down the street. Alternatively they may stop activity ; deer cease grazing when they hear a sudden, unusual sound. On plants the effects of stimuli are normally different from the effects on animals. The response of a plant is very much slower and usually consists of a growth movement, as when a shoot of a potted plant in a window grows towards the light.

In studying the effects of stimuli on organisms there are two basic questions to ask:
1 Exactly what stimulus has caused this particular change?
2 How does the response contribute to the survival of the organism?

Behaviour and survival

As we noted in Chapter 1, animals possess an advantage over plants in that they are mobile and consequently may be able to leave an unsuitable environment and seek a more favourable one. The more efficiently an animal can do this the more likely it is to survive in changing circumstances.

However, if behavioural mechanisms are to contribute to survival, the animal must be able to receive the appropriate stimuli from its environment and to assess and organize the information received, translating this into relevant activity.

Causal relationships

When we decide that a particular change in the environment is followed by a particular activity we usually imply a connection. A particular stimulus was presented and a particular response followed. This does not, of course prove cause and effect. Folklore provides numerous examples of the wrong interpretation of two concurrent phenomena. As Gilbert White wrote in 1767, 'In the autumn, I could not help being much amused with those myriads of the swallow kind which assemble in those parts. But what struck me most was, that, from the time they began to congregate, forsaking the chimnies and houses, they roosted every night in the osier beds of the aits of the river. Now this resorting towards that element, at that season of the year, seems to give some countenance to the northern opinion (strange as it is) of their retiring under water. A Swedish naturalist is so much persuaded of that fact, that he talks, in his Calendar of *Flora*, as familiarly of the swallow's going under water in the beginning of September, as he would of his poultry going to roost a little before sunset.'

Describing behaviour

'I heard a linnet courting
His lady in the spring:
His mates were idly sporting,
Nor stayed to hear him sing
His song of love . . .
        Robert Bridges.

For hundreds of years poets have praised the song of birds in spring, interpreting it as the bird's love song to his mate, or his hymn of joy at the return of the summer. To give it a technical name, the ascribing of human emotions to animals is known as *anthropomorphism*. We are not implying that the causes behind the behaviour of an animal are necessarily not the same as they would be in a man. Nevertheless, scientifically the anthropomorphic idea is only a hypothesis. To establish the hypothesis more firmly we would want to know what the animal was thinking, and there is, at present, no way in which we can do this. In other words, this sort of hypothesis is not yet capable of scientific test. Scientists must use hypotheses about animal behaviour which they can put to scientific test, and describe the behaviour that they see in terms as objective as possible.

In the case of the bird song, Eliot Howard found that the explanation in the case of at least one bird, the reed bunting, did not live up to the expectations of the poets. He showed that the bird's song probably had nothing *immediately* to do with the courtship. It was due to the male bird staking a claim to a particular area which he did before the female birds were in the area. He guarded this area–which Howard called the bird's territory–against other males. One way of doing this, apparently, was to sing and so warn other males that the territory was occupied. However, it is not only poets who anthropomorphize; biologists do it just as often and not so elegantly and we must take care that we do not do it unintentionally in our investigations.

Sense organs

The higher animals rely on complex sense organs to receive appropriate stimuli. The nature and complexity of these sense organs are of extreme importance in adapting an animal to a particular habitat. In birds, for example, visual information is of prime importance and the eye has developed as the most important sense organ. A bird of prey, depending upon sight to obtain food, may possess a visual ability of up to 20 times that of man. On the other hand, the dog depends largely upon smell, and its olfactory organs are much more sensitive than our own, although its sight is relatively poor. Man himself possesses no sense organ of extreme sensitivity. Although he obtains most of his information by sight, his other sense organs are sufficiently good to enable him to receive many sources of information from his environment.

The two sense organs which provide man with most information are the eye and the ear. How are light waves from an object transformed by the eye into information which can be assessed by the brain? How, similarly, are sound waves transformed by the ear?

Sense organs fall into the following four main categories, according to the type of information they receive:
1 Tactile receptor.
2 Chemical receptor.
3 Auditory receptor.
4 Light receptor.

However, it is not always necessary for an animal to possess a complex receptor system in order to respond to a particular stimulus.

Orientation

Animals orientate to some degree in relation to their environment. This is usually associated with locomotion, which is normally, but not always, directed. Animals normally move in response to certain stimuli and the result is usually that they reach a particular environment which bears a relation to the source of the orientating stimulus. Thus, as we have seen earlier (Chapter 1) *Tribolium* responds in a specific fashion when placed in a humidity gradient. In fact the distribution of animals is determined by their responses to various stimuli from their environment—humidity, light, temperature, chemicals, other animals, etc.

Reactions by which an animal orientates itself are useful means of studying the responses of animals to specific stimuli under laboratory conditions. Students of such reactions have developed a classification of the categories they have observed. This classification is useful in so far as it describes the behaviour observed. It must not be taken as a way of explaining the mechanism which determines the response. The two main groups of reactions of orientation are *kinesis* and *taxis*.

Kinesis

In this form of behaviour the animal's movements are not orientated with respect to the *source* of the stimulus. A change in the intensity of the stimulus results in a change in a quantitative aspect of the movement. For example, the animal may move more rapidly or turn more frequently.

Figure 28 shows that the locomotion of a planarian is affected by the intensity of light. When placed in strong light, the animal makes more turns per minute than in dim light. These turns are not related to the source of the stimulus.

Taxis

In this type of response the movements of the animal have a definite relationship to the source of the stimulus. In a positive taxis, the animal moves towards the stimulus and in a negative taxis, away from the stimulus.

Figure 28
The graph shows the effect of light on the locomotion of a planarian (*Dendrocoelum* sp.) as measured by the degrees of turning per minute. A to B was in darkness when the turning is at its lowest. At B a light was switched on and left on. The animal at first makes more turns per minute, the rate of turning then decreases as the animal becomes accustomed to the new conditions. The direction of turning is not related to the source of the stimulus.
From Ullyott, P. (1936) J. Exp. Biol., *13, pp. 253-278, 'The behaviour of* Dendocoelum lacteum'. *Cambridge University Press.*

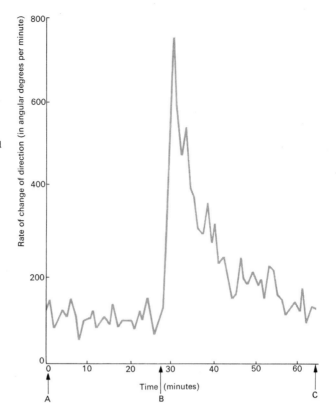

Orientation to light

The ability to make a response to a stimulus, such as light, will depend on the presence of a suitable receptor system. Apart from giving information about the direction of the stimulus, the receptors may also provide information about its intensity, the configuration and size of objects, and the distance of the source.

Figure 29
Larva of *Calliphora* sp.

The investigation is designed to study the type of reaction which an animal exhibits towards light, and its ability to detect differences in the intensity of light and to respond to them.

Calliphora larvae, the maggots of the bluebottle, are convenient animals to use for this purpose.

4.1 The reaction of *Calliphora* larvae to light

This investigation is intended to find out what type of reaction the animals make towards light, by following the track they make when exposed to it. Fully grown larvae, just before pupation, are the most suitable animals for experiment.

Procedure

1   Take a piece of black card, or other suitable material, about a foot square, and place it on the bench. Arrange a lamp in a suitable holder, just above one edge of the card and halfway along one side. With two retort stands and clamps, support a sheet of glass in a horizontal position about 5 cm above the surface of the card.

2   Black out the laboratory, and carry out the rest of the investigation with only the experimental lamps on. If there is another set of apparatus nearby, you may need to place a piece of black paper or card in position to cut out the light from other lamps.

3   First determine whether the animal's response to light is directional, by releasing one on the card near its centre. Follow its movements by marking the glass with a glass-marking pencil. When you think that you have detected a definite response, replace the animal in the centre and follow it again. Repeat this until you can be reasonably certain what is its response to light. Now try the experiment with other individuals and see if they respond in the same fashion.

4   It is possible to make a permanent record of the tracks the animals make by placing a sheet of paper on the top of the glass and tracing the marks. The glass sheet can then be cleaned in readiness for the next part of the investigation.

Questions

a   Did the animal's orientation to the light appear to be directional, or did the movement appear to be at random?

b   If there was a directional movement, was it towards the light *(photopositive)* or away from the light *(photonegative)*?

c   Is there any other factor which could have affected the response? If so, how could you try to eliminate its effects?

d   Did the individual respond each time in the same way? Did different individuals respond in a similar way? If not, was there any pattern to the variation or did it appear to be random?

4.2 Response of *Calliphora* larvae to varying light stimuli

An animal may be able to detect and respond to a directional light, but is it capable of responding to different intensities of directional light?

Procedure

1　Arrange the apparatus as before, but this time place the animals on the edge of the card, so as to give them the longest distance to move under the influence of the light. Place them on the side nearest the lamp if you think that they are photonegative, and furthest away if you think they are photopositive. If you are uncertain, then test your hypothesis by following the track of several animals orientating towards the light.

2　Now arrange a second lamp at right angles to the first about one-third to half way down one side. Ensure that the bulb is of the same strength. (If necessary test the strength of the two lights with a light meter at the points where the beams cross.) Draw a faint mark across the card to indicate the central position of the second beam.

3　With only the first lamp alight, start the animal off as before. When it reaches the line indicating the position of the second beam, switch on the second lamp and at the same time switch off the first. Follow the track of the animal orientating to the second light. Repeat this several times with the same animal and with other individuals.

4　Now try the experiment again, but this time leave the first lamp on when you switch on the second, and follow the tracks as before.

5　It should be possible, with a ruler and protractor, to measure roughly the angles that the animal's tracks are making in relation to the two lights. Try this and record all the results.

6　Now replace the second lamp with a more powerful one, or if this is not possible alter the position of the lamps so as to reduce the intensity of the first lamp (perhaps by putting a piece of paper tissue in front of the lamp). Repeat the experiment, following the tracks and measuring the angles.

Questions

a　How close to the direction of the first light beam did the animals keep? Remember that, unless you are using parallel rays of light from a ray box, the light rays will be from a diverging source.

b　How did the animals react to the second source of light?

c　When you had both lights on, to what extent were you able to predict where the animals would move? How far was your prediction correct?

d　What effect should an increase in the intensity of light from the second lamp have had on the orientation? Did it have this effect?

e　Supposing you had substituted a lamp of lower power from the second lamp, what effect do you think this would have had on the animal's response?

f　What is the importance of these responses to light in the normal life of the larvae?

4.3 Relationship of the response of *Calliphora* larvae to their structure

The light receptor of *Calliphora* larvae consists of a simple system. In fully developed larvae, similar to the type you have been using, the receptors consist of a number of light sensitive cells with nerves attached. These are situated, as shown in figure 30, in a pocket on either side of a thickened structure, the cephalo-pharyngeal skeleton, which forms the walls of a sucking pump used in feeding.

Procedure

1 Carefully examine the movement of the larvae when they are responding to a light source, paying particular attention to the anterior end where the receptors are located.

2 Record your description of the movements that occur at this end.

Questions

a Study the details of the receptor system given in figure 30. Can you produce a hypothesis to explain the movement you observed in the larvae and the action of the receptors?

b Can you devise any simple way of testing this hypothesis?

Figure 30
Diagram of light receptors of
Calliphora larvae.
After Bolwig, N., (1946) Senses and
sense organs of the anterior end of the
house fly larvae, *Vidensk. Medd.
Dansk*. naturh. Foren., Copenhagen.
109, 81–217.

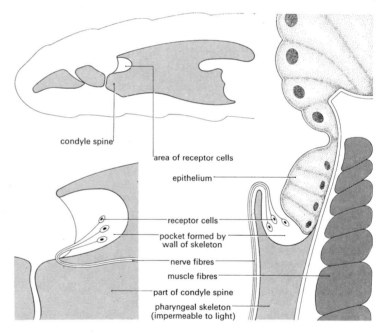

condyle spine

area of receptor cells

epithelium

receptor cells

pocket formed by
wall of skeleton

nerve fibres

muscle fibres

part of condyle spine

pharyngeal skeleton
(impermeable to light)

Plants and stimuli

The plant changes in many ways as the result of influences from its environment. For example, the size, shape, and number of leaves, flowers, and fruits and many other aspects of the form of a plant can be affected by environment.

Although woody plants cannot move from place to place, their non-woody shoots, petioles, and roots move a great deal even when environmental conditions are kept constant.

When changes occur, for example, in light intensity, movements may take place which can be related to the changed conditions. In one group of plant movements, these movements are connected not only with changing environmental conditions but also with specific changes in the direction of stimuli. These are *tropisms*; a *tropic* movement is one in which the direction of response is related to the direction of the stimulus. A tropism differs from a taxis in that only part of the organism shows movement in response to the stimulus.

Plant organs may respond directionally to light, gravity, water, and physical contact but the work of this chapter is concerned only with two factors: the response to light *(phototropism)* and the response to gravity *(geotropism)*.

Although stems, petioles, and roots of plants exhibit tropic responses, these are most evident in the shoots and roots of seedlings. Root tips have a simple, regular outline but the tips of plumules are complicated. Monocotyledon plants, like grasses and cereals, have their first leaves enclosed in a cylindrical sheath called a *coleoptile* and because these are simple and regular in shape most of the work on phototropism has been done using coleoptiles of oat *(Avena)* and wheat *(Triticum)*. Barley *(Hordeum)* is a possible alternative plant.

4.4 Structure of coleoptiles

Procedure

1   Take a germinating oat or barley grain which has an intact coleoptile about 2–3 cm long. Cut it at the base, near the grain, and remove the coleoptile. Cut the coleoptile longitudinally with fine scissors. Unroll it and mount a small piece in glycerol or water under a cover-slip.

2   Do the same with another piece but cover it with 2–3 drops of phloroglucinol solution. Leave for 3–4 minutes and then replace the phloroglucinol with a drop of concentrated hydrochloric acid. Mount under a cover-slip.

3   Examine your preparation under low and high power magnification.

Questions

a Is the coleoptile tissue homogeneous as seen by the naked eye or hand lens?

b How many different kinds of cells can be seen under low or high power magnification?

c Describe the shape and proportions of the commonest type of cell.

d Phloroglucinol and hydrochloric acid cause lignin to turn red. Is there any lignin present in the piece of coleoptile?

4.5 The effect of unidirectional light on a coleoptile

Germinating oats or barley grains with intact straight coleoptiles are suitable material for this investigation. It is concerned with phototropism. This is said to be positive or negative according to whether the plant organ moves towards or away from the source of the stimulus.

Procedure

1 Fit a graticule (linear scale) into the eyepiece of a microscope. Bend a piece of wire or thin sheet metal into the shape shown in figure 31. Push this gently into position on top of the graticule so that it is held firmly in place when the eyepiece is held horizontally. Insert the eyepiece into a microscope which has been arranged horizontally.

2 If the laboratory has windows in only one wall, set the microscope parallel to this wall, i.e. at right angles to the path of light. If daylight strikes the microscope from several directions, shield it by erecting screens and place a projector or microscope lamp to one side of the microscope stage.

3 Take an oat or barley grain which has an intact straight coleoptile (i.e. with no leaves protruding) and set it in a mount similar to that shown in figure 32. Record the way in which the coleoptile has been grown, (in the light or in darkness).

Figure 31
A means of fixing a graticule.

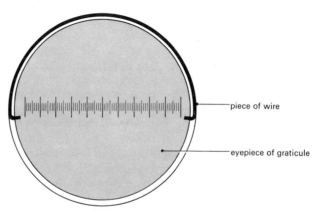

piece of wire

eyepiece of graticule

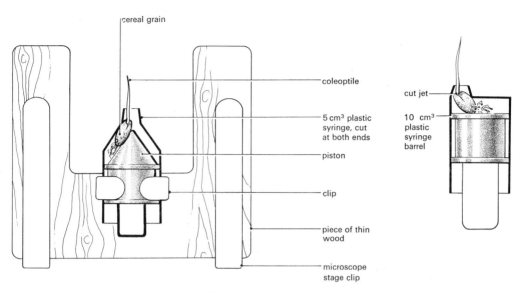

Figure 32
A device for holding a coleoptile on a
microscope stage.

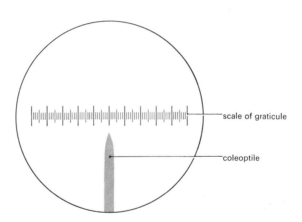

Figure 33
A coleoptile or radicle tip and a
graticule scale.

4   Focus the low power objective (×3 to ×5) on the coleoptile
and move the mounting device about until the coleoptile tip
appears in contact with the mid-point of the graticule scale
(5·0). Set the scale horizontally. (See figure 33.) Note the
time ; start a stop-clock.

5   After two minutes note the position of the tip of the coleoptile
against the scale. Make slight adjustments of the graticule,
if necessary, so that the coleoptile tip appears to move along
the scale rather than through it. Make further readings of
the coleoptile position at 1–2 minute intervals until a definite
and regular movement has been recorded. This should take
5–8 minutes.

6   Note the time and rotate the mounted coleoptile through 180° on its vertical axis. Reset the tip of the coleoptile on the mid point of the graticule scale. Continue, as before, to take readings of the coleoptile position at 2–5 minute intervals for 1 hour.

7   Plot the successive positions recorded, as a graph. Remove the mounted coleoptile, examine it, and draw a line representing its final shape in relation to the position of the light source.

Questions

a   Does the coleoptile tip initially move towards or away from the light source?

b   When the coleoptile was rotated through 180° the direction of light was, in effect, reversed; was the movement of the coleoptile immediately reversed?

c   If not immediately, after what period of delay?

d   Did the whole coleoptile bend, or was there a definite region which reacted?

e   In what way is this response likely to benefit the plant's growth under natural conditions?

f   How can you account for your experimental results? Give the simplest explanation which could serve as a working hypothesis for further experiment.

4.6 Structure of radicles

The roots of cereal plants, such as oats or barley, are unsuitable for investigations into the effects of gravity. Germinated mung beans provide a suitable alternative and can also be used to investigate the structure of radicles.

Procedure

1   Take a germinated mung bean (or pea) with a radicle 1–2 cm long and examine it carefully with a hand lens.

2   Mount a piece of radicle in water under a cover-slip and examine some root hairs under a high power objective.

Questions

a   Where is the root cap located and what is its length?

b   What is the position and length of the region bearing root hairs?

4.7 The effect of gravity on a radicle

Geotropism, like phototropism, can be described as positive or negative.

Procedure

1   The radicle of a mung bean is of similar size to a barley coleoptile. Select a bean with a fairly straight radicle 1–2 cm long. This should be fitted into the device previously used (4.5). To prevent desiccation enclose the syringe and the radicle with a suitable cover, such as a boiling tube.

2   Fix a graticule in an eyepiece as shown in figure 31 and described above (4.5). Arrange a microscope in a horizontal position and fix the mounted radicle so that it also lies in a horizontal position. Rotate the eyepiece until the graticule scale is vertical and move the radicle until its tip appears to touch the mid-point of the scale (5·0).

3   Start a stop-clock and record the position of the radicle tip at 1–2 minute intervals for 10 minutes. After a known period (10–20 minutes), rotate the mounted radicle through 180° about the horizontal axis. Reset it so that the tip is again touching the mid-point of the scale.

4   Continue to take readings at 2–5 minute intervals for about 1 hour. Plot your readings as a graph. At the end of the experiment remove and examine the radicle. Make a note of its final shape.

Questions

a   What evidence have you obtained to counter the suggestion that a radicle merely sags, like a non-living structure, under the influence of gravity?

b   Did the whole radicle bend, or was there a definite region which reacted?

c   If there is a definite bend in the radicle, where is it in relation to the root cap and the root hair region?

d   What advantage would this response bring to a plant growing in its natural habitat? Would you expect all roots from the plant to behave in a similar fashion?

e   Suggest a hypothesis to account for the positions of the radicle you recorded during the 1 hour period, including the fact that the radicle was rotated through 180°.

Tropic responses and growth substances

Many experiments involving the effect of a unidirectional stimulus, such as light, have provided evidence that the observed effect is due to the action of a chemical agent. Further information about the importance and use of growth substances of this nature is to be found in the Topic Review *Auxins and their applications*.

As plants lack muscle tissue, one explanation for tropic movements could be that they are achieved by changes in cell size. Earlier we saw that changes in the size of cells can produce changes in the shape and strength of plant tissues.

4.8 Effect of IAA on the growth of coleoptiles and radicles

Chemical agents, auxins, which affect the responses of plants to stimuli have been extracted from plant tissue and identified. One of the commonest is indole-3-acetic acid (IAA). Does this affect the size of isolated coleoptiles and radicles?

Procedure

1   Prepare five Petri dishes, each containing one of a range of IAA and sucrose solutions, as follows:

1 For the first, take 10 cm³ of stock IAA solution (0·2g/dm³ IAA) and add 10 cm³ of 4 per cent sucrose solution.

2 For the second, take 1 cm³ of stock IAA and make up to 100 cm³ with water. To 10 cm³ of this solution add 10 cm³ of 4 per cent sucrose.

3 For the third, take 1 cm³ of this diluted stock and make up to 1 dm³ with water. Use 10 cm³ of this with 10 cm³ of 4 per cent sucrose.

4 For the fourth, take 1 cm³ of the twice diluted stock and make up to 100 cm³ with water. Mix 10 cm³ of this with 10 cm³ of 4 per cent sucrose solution.

5 For the fifth, add 10 cm³ of water to 10 cm³ of 4 per cent sucrose solution. This is the control.

You should now have 5 dishes with the following contents:

1 100 parts per million IAA + 2 per cent sucrose.

2 1 p.p.m. IAA + 2 per cent sucrose.

3 10^{-3} p.p.m. IAA + 2 per cent sucrose.

4 10^{-5} p.p.m. IAA + 2 per cent sucrose.

5 2 per cent sucrose.

2   Cut 10 mm sections from 25 coleoptiles and place 5 in each dish. To do this, place some coleoptiles in a row and line up their tips. Then cut the sections starting 1–2 mm back from the tip. All the sections must be exactly the same length (10 mm); the best way of achieving this is by first making a simple cutter from two razor blades. See figure 34a.

Figure 34
Ways of cutting 10 mm sections.

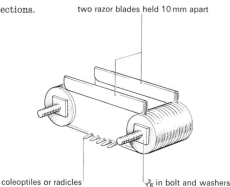

two razor blades held 10 mm apart

coleoptiles or radicles

$\frac{3}{16}$ in bolt and washers

a

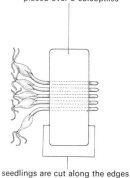

a piece of metal exactly 10 mm wide placed over 5 coleoptiles

seedlings are cut along the edges

b

Alternatively, a simpler but less effective method is to line up the ends of the coleoptiles first and then hold them down with a strip of metal, wood, or celluloid exactly 10 mm wide, and make two cuts with a very sharp scalpel or new razor blade. See figure 34b.

Whichever method you use, the cuts should be made at right angles to the long axis of the coleoptiles and radicles.

Ideally, you should cut and transfer the coleoptiles into dishes in a darkroom illuminated only by red light. If this is not possible, carry out these operations quickly in as dark a place as you can find.

3 Put lids on the Petri dishes and place them in a dark incubator set at 25° C.

4 After 48 hours remove the dishes and measure the length of each coleoptile and radicle section to the nearest 0·5 mm. You can do this by placing each section alongside an ordinary ruler and observing both through a hand lens.

Record your measurements:

Set no.	Length of coleoptile section	Length of radicle section
	1.	1.
	2.	2.
Solution containing	3.	3.
.... p.p.m. IAA + 2 per cent sucrose	4.	4.
	5.	5.

Questions

a Obtain measurements of coleoptile and radicle sections made by others in the class. Calculate the average for each treatment (there are five treatments for coleoptiles and five for radicles). Subtract 10 from each average to express the change from the original length (10 mm).

b You may have assumed, at the outset, that all the coleoptiles or radicles in any one solution would respond in exactly the same way. There are two reasons why response may not appear uniform. These are

possible errors of measurements, or

possible differences in metabolism between the sections producing genuine differences in response.

If you have sections of different lengths, but treated in the same manner, the obvious step is to calculate averages for each experiment.

An average is not much use unless you know how typical it is of the individual results, i.e. unless you know the degree to which they vary from the average.

You can obtain some measure of the reliability of averages by calculating the standard deviation. Do this for each of the 10 experiments, as follows:

1 Calculate the deviation (d) of each measurement from the average for that experiment (reading minus Av. = d).

2 Square each deviation (d²).

3 Add all the squared deviations (Σ d²).

4 Obtain the standard deviation (S) by using the formula

$$S = \sqrt{\frac{\Sigma\, d^2}{N-1}}$$ where N = the number of results from any one experiment.

The larger the standard deviation the greater is the error of the estimate made. The average plus or minus the standard deviation will include about two-thirds of the cases observed in a normally distributed collection of results.

c   Plot a graph of the results obtained. Let each unit on the horizontal axis represent an increase of IAA concentration × 10. Against this, mark not only the average lengths obtained in each treatment but also each standard deviation.

d   Does the application of IAA appear to have a definite effect on coleoptile sections and radicle sections? If so, is the effect roughly proportional to the concentration of IAA supplied? Do coleoptile and radicle sections respond in the same way to the same concentration of IAA?

e   Some concentrations of IAA may have a very marked effect on the sections. However, with other concentrations there may be some doubt because there is only a small difference between the growth of sections in one concentration of IAA and those in treatment *(5)* which contained no IAA. You have to decide whether the differences are due to chance or to some biological effect.

There are several ways of doing this statistically. One useful method is by calculating the standard error of the averages of the two experimental results concerned.

Suppose you are not sure if there is any significant difference between the results of treatment *(1)* (100 p.p.m. IAA) and treatment *(5)*, the control (no IAA). You will have calculated the average values for coleoptiles as answers to question (a), and the standard deviations as answers to question (b). Calculate the standard error (SE) of the difference:

$$\text{SE average 1} - \text{average 5} = \sqrt{\frac{S^2_1}{N_1} + \frac{S^2_5}{N_5}}$$

Where S_1 = standard deviation of results from treatment *(1)*
S_5 = standard deviation of results from treatment *(5)*
N_1 = the number of results obtained in treatment *(1)*
N_5 = the number of results obtained in treatment *(5)*.

If the difference between the two averages (average 1 — average 5) is larger than *twice* the standard error of the difference, it may be concluded that the difference is not due to chance but to a genuine biological response to the treatments.

Apply this method when comparing the results of treatment *(5)* (control) with others in cases where the differences between averages are small.

f   In what ways do the results obtained with IAA aid the understanding of tropic responses?

State clearly the connection between growth, as measured in these experiments, with tropic responses observed earlier, and the possible part played by IAA in both phenomena.

Bibliography

Bishop, O.N. (1966) *Statistics for biology.* Longman. (Standard deviation and other statistical methods.)

Carthy, J. D. (1966) *The study of behaviour.* Edward Arnold. (Chapter on orientation behaviour.)

Fraenkel, G. S. and Gunn D. L. (1961) *The orientation of animals.* Dover. (Classic work on orientation from which many ideas for further work can be obtained.)

Sands, M.K. (1970) Nuffield Advanced Biological Science Topic Review *Auxins and their applications.* Penguin.

Simon, E. W., Dorner, K. J., and Hartshorne, J. N. (1966) *Lowson's textbook of botany.* 14th edition. University Tutorial Press. (Account of auxins.)

Strafford, G. A. (1965) *Essentials of plant physiology.* Heinemann Educational Books. (Extended account of plant hormones.)

White, Rev. G. (1767) *The natural history of Selborne.*

Synopsis

1 The nervous system brings about the co-ordination of muscular movement essential for efficient behaviour.

2 By dissection, the anatomical relation between nerves and the tissues they supply can be found.

3 Electrical stimuli cause nerve impulses in motor nerves which stimulate the contraction of skeletal muscle.

4 Different patterns of stimuli produce different effects on muscles.

Chapter 5

Nerves and movement

Summary of practical work

section   *topic*

5.1   Distribution of nervous tissue in the hind leg of a frog

5.2   Histology of nerve tissue

5.3   Dissection of a frog sciatic nerve/gastrocnemius muscle preparation

5.4   The nature of muscle twitch

5.5   Effect of two successive stimuli applied to the gastro-cnemius preparation

5.6   Effects of repetitive stimuli

Nervous system

Animals are mobile and are capable of responding rapidly to changes in their environment. As we have seen, their behaviour has considerable survival value.

Much, but not all, behaviour involves the animal in movement. Except for anatomically simple animals such as protozoans, most animals possess a nervous system and it is through this that the movement is co-ordinated.

5.1 Distribution of nervous tissue in the hind leg of a frog

By careful dissection we can follow nerves from the point where they leave the central nervous system to the structures which they innervate. The sciatic nerve of a frog is particularly appropriate for this dissection as we shall use the same nerve in investigation 5.3.

Procedure

1 Pin a freshly killed frog with the ventral side down in a dissecting dish. Cover the animal with water and make a median incision through the skin of the back, then carefully pin out the freed skin. Note the dorsal cutaneous nerves lying free in the lymph spaces of the back.

2 Cut through the muscles of the back along the mid-line. The incision should extend from the middle region of the vertebral column to the tip of the urostyle. Fold the freed muscles of one side outwards and look for spinal nerves 7, 8, 9, and 10, which join to form the sciatic plexus.

3 Make a median incision through the skin of the leg as far as the foot. Pin out the skin, taking care not to break any nerves associated with it. Locate the white tendon which is centrally placed in the thigh region, and separate the muscles underneath it by blunt dissection to display the sciatic nerve.

4 Trace the sciatic nerve from the plexus into the thigh by separating the muscles of the thigh and pinning them to display the nerve. Trace as many branches of the nerve as possible as far as you can. Continue the dissection into the rest of the leg and foot.

5 Make a sketch or a diagram of your dissection. Annotate it to show the parts of the leg to which you have traced the finest branches of the sciatic nerve.

Figure 35
Galvani's experiment.
Photo, Science Museum, London.

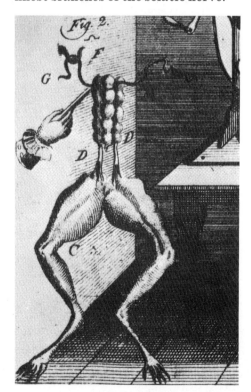

Questions

a What functions are associated with the parts of the leg supplied by the sciatic nerve?

b Would you expect the nerves supplying these structures to be conducting information from a receptor, or to an effector?

c Check the arrangement of the sciatic nerve system from the origin of the spinal nerves to the ends of the sciatic nerves in the leg in different specimens. How much individual variation do you find? In what part does the variation occur? Is this likely to have any functional effect?

5.2 Histology of nerve tissue

The basic function of a nerve is to conduct information in the form of nerve impulses from one part of the body to another. Can we see any details of structure in the nerve fibres which will throw light on how it functions? It is best to use permanent preparations as these provide a clearer picture than fresh material.

Procedure

1 Examine, under low power, permanent preparations of teased nerve and transverse and longitudinal sections of nerve.

2 When you have seen the general structure of nerve, examine it under high power to determine the finer details.

3 Answer the questions and make annotated sketches or diagrams to illustrate the structure of a nerve and its component nerve fibres.

Questions

a What component of nervous tissue forms the major part of a nerve? What connective tissue supports this major component?

b Are the nerve fibres of the same or different sizes? Do the individual fibres possess any form of covering? If so, what form does this covering take?

c Can you identify any cell nuclei? If so, what is their relationship with the individual nerve fibres and with any covering they may possess?

d From an examination of osmic acid preparations what can you say about the chemical nature of the outer layer of a nerve fibre?

e Which part of a nerve fibre do you consider is responsible for the conduction of nerve impulses?
Is this part made of many cells, one cell, or a part of a cell? Can this part, be traced to a nucleated cell body in a nerve?

f Do the nerve fibres join to form a network or do they appear to retain their individuality; or is it impossible to decide from an examination of the slides provided?

Nerve and muscle

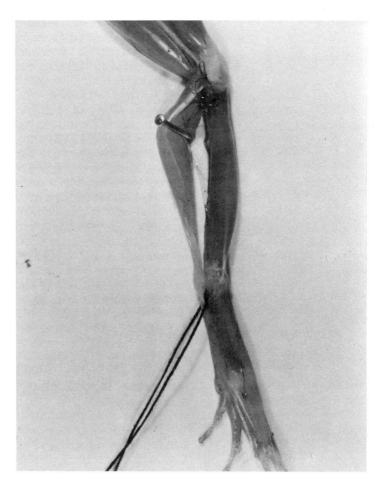

Figure 36
Dissection of the frog. Gastrocnemius
muscle exposed with thread ready to
tie around Achilles tendon.

Galvani, who was a professor of anatomy at Bologna, made
his celebrated discovery of nerve action by chance. In his
own words, 'The discovery was made in this way. I prepared
a frog as represented [in figure 35] and while I was attending
to something else, I laid it on a table on which stood an
electric machine at some distance from its conductor and
separated from it by a considerable distance. Now when one
of the persons who were present touched, accidentally and
lightly, the inner crural nerves DD of the frog with the point
of a scalpel all the muscles of the legs seemed to contract
again and again as if they were affected by powerful cramps.'

This discovery would appear to have been made about 1771.
For the next twenty years Galvani investigated the pheno-
menon, vigorously clinging throughout to the belief that
electricity had been generated by the frog and that the metal
of the scalpel merely allowed this animal electricity to flow
into the muscle from the nerve. One of the ideas that he

investigated experimentally was that lightning might evoke the same response. He prepared frogs for exposure to lightning by driving iron hooks through their backbones and spinal cords. He then suspended the frogs by the hooks in a horizontal line above an iron grating in his garden. One of the greatest discoveries in science was made on a September day in 1786 when there were no thunderstorms. Instead of lightning the wind played on the frogs, jogging them up and down so that the hooks came into contact with the iron railing. Every time this happened the frogs twitched. The frogs, acting as sensitive instruments, had detected the electric current generated by the contact of the two metals. Although both metals were iron, slight physical or chemical differences between the hooks and the railing would produce an electric potential sufficient to excite the muscles. Although Galvani stated that the hooks were iron, many textbooks erroneously say they were brass or copper. This mistake is insignificant compared with the one Galvani made in interpreting the results of his experiment. He completely missed the true explanation, and in a paper published in 1791 he stuck to his original view that electricity had been generated by the animal and the contact of the metals had merely provided a conducting pathway.

The Italian physicist, Volta, disagreed with Galvani. In a letter to the Royal Society, Volta gave the correct explanation, that the electricity generated between the hooks and the railings had stimulated the nerves of the dead animals from the outside. When we call the electricity generated between dissimilar metals galvanic, we pay tribute to the man who unwittingly discovered it. Volta went on to construct a voltaic pile consisting of several plates of copper and zinc separated by cloths soaked in acid or salt solution. He compared such batteries to the stack of plates in electric fish and rightly deduced that they produce electricity by similar means. The work of Galvani and Volta laid the foundations of the study of both electricity and neurophysiology.

Since this time, biologists owe much of what they have come to understand about the physiology of nerves and muscle contraction by using a nerve/muscle preparation. The nerve of such a preparation can be stimulated to conduct impulses and cause the muscle to contract. These contractions can be recorded by use of a kymograph. It is more convenient to use a cold-blooded animal because preparations from warm-blooded animals require more complicated equipment. This is because it is necessary to maintain the animal's body temperature and also to aerate the bathing liquid continuously. The most convenient way to stimulate the nerve is to give it an electric shock. For this purpose we can use a piece of apparatus called a stimulator. This will normally provide electric shocks over a range of voltages and a variable number of repetitive shocks per second.

5.3 Dissection of a frog sciatic nerve/gastrocnemius muscle preparation

Procedure   1   Cut the skin around the middle of the body of a frog, freshly killed by pithing. By using fingers and blunt forceps pull the skin backwards and with sharp tugs pull it off the hind legs. Throughout the dissection keep the animal moist with Ringer's solution.

Figure 37
Dissection of the frog. Sciatic nerve gastrocnemius muscle preparation, with nerve fully exposed.

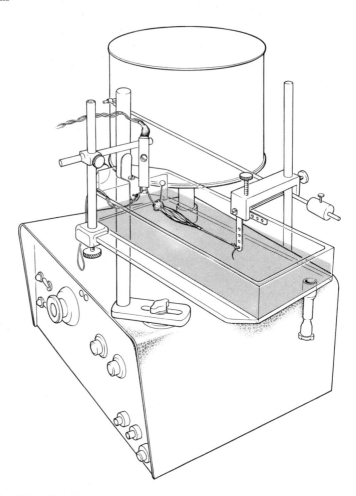

Figure 38
The muscle bath and kymograph.

2   Identify the gastrocnemius muscle and Achilles tendon in
    one leg. Using a mounted needle separate the tendon from
    connective tissue and slip a piece of cotton thread behind
    the tendon. Tie the thread just anterior to the sesamoid
    bone. Cut the tendon posterior to the sesamoid bone and
    free the gastrocnemius muscle from the tibio-fibula bone
    leaving the muscle attached only at the knee region.
3   Place the frog with its dorsal side uppermost. Using wooden
    or plastic cocktail sticks separate the muscles along the
    back of the thigh and locate the sciatic nerve. Avoid touch-
    ing this with any metal instrument, then carefully trace it
    back as far as you can, and forward to the knee joint. At the
    anterior end slip a loop of cotton thread behind the sciatic
    nerve and make a knot. Cut the nerve anterior to the liga-
    ture. Then, still using the cocktail sticks, carefully separate
    the nerve from the muscle as far as the knee joint.

4   Cut the femur not more than 1 cm from the knee joint and the tibio-fibula just below the knee joint. Insert a pin through the knee capsule.

5   Place the preparation in a muscle bath containing Ringer's solution. To secure the preparation push the pin into the cork base or the hole, if there is one. Next tie the thread which is attached to the tendon, to the lever. Adjust the support of the lever until the lever itself is horizontal. By use of the thread attached to the nerve, lift the nerve out of the solution and place it over the stimulating electrodes. The completed set-up should look like figure 38. During subsequent investigations ensure that the nerve is kept continuously moist with solution.

5.4 The nature of muscle twitch

Procedure

1   To set up a kymograph, first remove the drum. Place a sheet of recording paper around it and smooth the paper down. After ensuring that you have placed the paper so that the edges overlap in the same direction as the recording pen moves, stick the edge down. Replace the drum on the spindle.

2   If you are using an ink pen, check that this is writing smoothly by moving the muscle chamber so that the pen is in contact with the paper on the recording drum. Turn the drum by hand and check that the pen is producing an even ink line. If you are using a smoked drum, check, similarly, that the lever is marking evenly. Switch on the stimulator, set the voltage control to its lowest point, and deliver one stimulus. Turn the drum about 2 cm. Increase the voltage and repeat. Continue this until you find the strength of stimulus required to give precisely a maximum response. Maintain the stimulator at this voltage for all subsequent experiments on the preparation.

3   Kymographs normally possess a switching arrangement so that a stimulus can be delivered at the same point of the drum surface as it rotates. The simple circuit for such an arrangement is shown in figure 39. Bring this switch into action so that one shock will be given for each complete drum rotation.

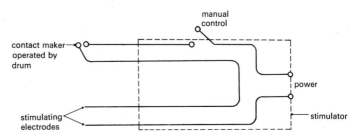

Figure 39
Circuit for the delivery of drum-triggered stimuli.

4   Arrange the drum on the spindle so that the recording will be made about halfway down. It may then be possible to make a second trace below the first. Set the switch on the kymograph so that it can be most easily run at maximum speed. Move the pen slightly away from the drum, switch on the kymograph, and allow it to revolve a few times to gain maximum speed.

5   Bring up the pen so as to touch the paper surface and produce an even line. Operate the control to allow one stimulus to be delivered. Immediately the response has been made, switch off the kymograph.

6   To mark the position on the drum where the stimulus was delivered, disengage the motor and turn the drum by hand until the kymograph switch is about to be closed, operate the stimulator control, and turn the drum slowly until the stimulus is delivered and the response recorded on the drum.

7   Move the pen away from the drum and adjust the drum on its spindle so as to leave room for another recording below the first one. Repeat the procedure to obtain the response to a single stimulus as before.

8   Gently lift the nerve off the electrodes and lower it into the Ringer's solution. Arrange the electrodes on the surface of the muscle in such a position that the muscle may be electrically stimulated and yet be free to contract. Repeat the procedures to give the muscle a single stimulus by direct contact with the electrodes.

Questions

a   During the course of the dissection, did the muscle show any signs of being stimulated? If so what do you think caused the stimulation?

b   What events are recorded when a single stimulus is delivered to the nerve to produce a muscle twitch?

c   How do you account for the muscle giving a graded response to stimuli of different strengths? Why, after a certain voltage has been reached, does the muscle no longer give an increased response to an increased stimulus?

d   What is the time period between a stimulus being delivered and the muscle responding? What mechanical and physiological factors could contribute to this?

e   What is the time delay when the muscle is stimulated directly? What can be deduced from this about the time delay due to conduction along the nerve?

f   What opposes the shortening of the muscle fibres? How do natural contractions of the muscle differ from these experimental twitches?

Successive stimuli

We have seen the effect of a single stimulus given to a muscle under experimental conditions. However, under normal conditions it might well be that more than one impulse reaches the muscle in a short time period via the motor nerve. What would be the effect of two impulses reaching the muscle with a short time lapse between them?

5.5 Effect of two successive stimuli applied to the gastrocnemius muscle preparation

Procedure

1   Use the same preparation as for 5.4 or set up a similar one. Bring the other contact switch on the kymograph into action so that the two switches are separated by about 30° to 40°.
2   Set the kymograph to operate at maximum speed as before. Set the stimulator to produce precisely the maximum response.
3   Switch on the kymograph and record the response to the two stimuli. Stop the kymograph and mark the record with the angle between the two contact switches.
4   Bring the two contact switches slightly closer together and repeat the procedure. Continue this until the contact switches are so close together that the effect is similar to that obtained by one switch alone.
5   Convert the angles of separation into time intervals and add these to the record.

Questions

a   What effects do two successive stimuli have on muscle twitches?
b   At what time intervals were effects shown?
c   What explanation can you give for the change in effect when the two successive stimuli are given with a very short time period between them?
d   What method of grading muscle activity does this investigation demonstrate?

Summation and tetany

The interaction of two stimuli on the contraction of a muscle is called *summation*. If a larger number of successive stimuli is given, the muscle will sustain a state of maximum contraction called *tetany*. The second wave of shortening develops greater tension than the first because some of the initial shortening is expended in taking up the slack in the system. The additive effect of the two contractions, one on top of the other, gives summation of mechanical response.

Infection by tetanus bacilli can result in a disease called lock-jaw in which the muscles closing the jaw go into a state of tetany. Strychnine poisoning produces an analogous effect and causes tetany of various muscles in the body.

5.6 Effects of repetitive stimuli

Procedure

1   Again, use the preparation as in 5.4 or a similar one. Set the kymograph to operate at one revolution per minute. Complete the circuit for stimuli to be delivered by operating the control on the stimulator alone. Adjust the strength, as before, to just produce a maximal response.

2   The purpose of the experiment you are to make is to record the effect of increasing the number of stimuli given per second. Three separate recordings should be made in the course of one rotation of the drum. One set should record a frequency of 5 stimuli per second, the next 10 per second, and the third 20 per second.
Stimulate the preparation for only three seconds on each occasion. You must allow time between the volleys for the muscle to relax, so stop the kymograph and allow the preparation a rest period of about 20 seconds between each set of stimuli.

3   Mark each recording with the frequency of the stimulation. Remove the paper and measure the amplitude of each set of recordings and compare these with the peak of a single twitch.

Questions

a   How does the extent of contraction vary with the frequency of stimulation and at what frequency did maximum contraction (tetany) occur?

b   How does the extent of contraction at tetany compare with the peak of contraction of a single twitch evoked by a suitable stimulus?

c   From the information you have obtained on the musculature and innervation of the frog hind limb, try to assess how far the observed movements can be explained in simple terms of single muscles contracting.

d   Observe the hind limb movements of a living frog on land and in the water. What effect will contraction of the gastrocnemius muscle have on limb movements?

e   What is the effect of exposing the muscle preparation to continual stimulation? How could you investigate the effect of continual muscle contractions of a human muscle, such as the one which bends a particular finger joint?

Bibliography

Clegg, A. G. and Clegg, P. C. (1962) *Biology of the mammal.* Heinemann Medical Books. (Account of nerve impulse and the brain.)
Freeman, W. H. and Bracegirdle, B. (1966) *An atlas of histology.* Heinemann Educational Books. (Full histological study of nerve tissues.)
Marshall, P. T. and Hughes, G. M. (1965) *Vertebrate physiology.* Cambridge University Press. (Chapter on nervous co-ordination.)

Chapter 6

Structure and function in the nervous system

Summary of practical work

section   *topic*

6.1   The roots of spinal nerves

6.2   Regions of the central nervous system

6.3   Complexity of brain structure

6.4   Use of a cathode ray oscilloscope to study the nerve responses in a frog sciatic nerve

6.5   Responses in the locust's central nervous system

Central nervous system

Apart from a few simple animals, all animals with a distinct nervous system have concentrations of nervous tissue forming co-ordinating centres. In vertebrates these form the central nervous system, composed of the brain and spinal cord. Nerve impulses from receptors pass along sensory nerve fibres to the central nervous system from which impulses leave along motor fibres to effectors.

Reflex action

In a vertebrate such as man, or a frog, the simplest type of action initiated by the nervous system is known as a 'reflex action'. This originally implied that nerve impulses travelling in sensory nerves were 'reflected' and passed into motor nerves.

Stephen Hales (1677–1761), a vicar of Teddington, near London, spent much time, for a number of years, investigating biological phenomena. He was one of the first investigators to bring a quantitative rather than a purely descriptive approach to his subject. He observed that a decapitated frog will still respond to certain stimuli. If the preparation is suspended, the legs hang limply down, but, if the foot is stimulated by pinching, then the leg is drawn up.

However, he did not establish the significance of the spinal cord in this process. This was left to a Professor of Medicine at Edinburgh University, Robert Whytt (1714–1766). He was able to demonstrate experimentally that such a reflex action depended upon the presence of the spinal cord, although an intact portion of it would suffice. 'When the body of a frog is divided into two, both the anterior and posterior extremities preserve life and power of motion for a considerable time.' This work provided a basis for the later studies of neurophysiology in the nineteenth century which established the mechanism by which these reflex actions occur.

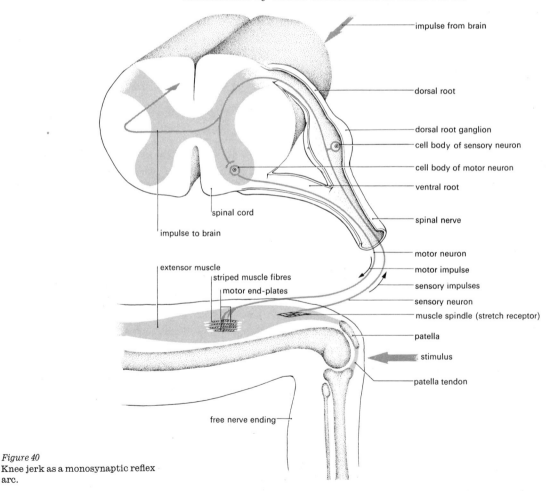

Figure 40
Knee jerk as a monosynaptic reflex arc.

Mechanism of a reflex action

One of the simplest types of reflex action known is the knee jerk, frequently used as a diagnostic test by doctors. The hypothetical pathway of this action shown in figure 40.

However, even this must be more complex than the diagram suggests. Even when the eyes are closed, you know that the knee has been tapped and the leg has moved. For this to occur, information must have been transmitted to the brain.

Many reflexes involve the brain and not the spinal cord. A touch on the cornea of the eye, for example, produces an automatic blink.

6.1 The roots of spinal nerves

Nerve fibres enter and leave the spinal cord only in the spinal nerves. Their anatomical relations can be investigated to locate any structural basis for their functioning. We can discover the origin of a spinal nerve from the spinal cord by dissecting a suitable animal, such as a frog. It is best to do this experiment on a specimen which has been left in an acid solution for some time, to decalcify and soften the bones.

The best way of studying the micro-anatomy of spinal nerves is to use prepared microscope slides, suitably stained.

Procedure

1   Cut through the vertebrae dorsally and carefully remove the roof of the neural canal to expose the spinal cord.
2   Trace one spinal nerve into the neural canal and identify its roots with the spinal cord. Check that other spinal nerves have similar roots.

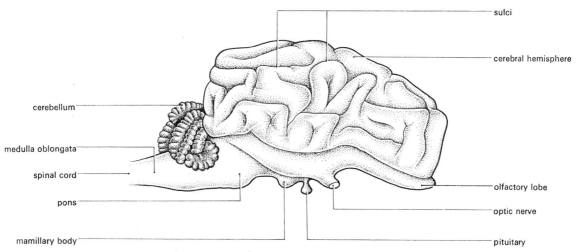

Figure 41
Diagram of a sheep's brain.

3   Make annotated diagrams or sketches of the roots. Show the position of any ganglia (swellings) you observe.
4   Examine under low power a permanent microscope preparation in which a dorsal and ventral root have been cut longitudinally. Identify the two roots.
5   Examine the slide under high power. Identify the components of nervous tissue present in each root. Make a labelled diagram or sketch of the preparation.

Questions

a   How can the dorsal root be distinguished from the ventral root by its gross appearance?
b   In which of the roots can the cell bodies of neurons be seen?
c   What is the maximum number of fibres (axons) that can be seen extending from the cell bodies?
d   How is the cell nucleus orientated with regard to the rest of the neuron?
e   How far can you decide upon the functions of the two spinal roots from their structure?

6.2 Regions of the central nervous system

Investigations into reflex actions such as the withdrawal reflex have shown that they only occur in an animal with an intact spinal cord. Information is relayed to and from the brain via the spinal cord. So far we have identified two components present in nervous tissue–nerve fibres and nerve cell (neuron) bodies. We can follow the distribution of these components by the use of a differential stain.

Procedure

1   The tissue will have been fixed in formalin. Cut the tissue into slices and wash them thoroughly in running water. It is convenient to cut the spinal cord into transverse slices. It is best to cut the brain into longitudinal slices about 8 mm thick.
2   Place the slices in Mulligan's fluid for two minutes. Then wash them gently in running water for one minute, rocking the container to keep the slides moving.
3   Place the slices in tannic acid for one minute. Then wash them gently in running water, as before, for three minutes.
4   Place the slices in 1 per cent aqueous iron alum until the original grey portions of the tissue turn black. This may take 30 seconds or less. Wash well.
5   Examine a permanent prepared transverse section of spinal cord under low and high power. Find good examples of neuron cell bodies under high power and make sketches or diagrams to illustrate their cytological structure.

Questions

a   What are the positions of the darkened areas present in slices of spinal cord and brain tissue?

b   From your microscopical examination of the spinal cord can you correlate the distribution of nerve fibres or neuron cell bodies with the darkened areas?

c   What hypothesis could you make about the distribution of fibres and cell bodies in the brain?

d   What are the distinctive features of the cell bodies in the spinal cord? How can they be distinguished from those found in a spinal root?

e   How far would you be justified, on the results of these investigations alone, in assuming that the fibres of nerves are processes extending from neuron cell bodies in the spinal cord or spinal root?

6.3 Complexity of brain structure

Sherrington described the brain, in which there are more cells than there are people in the world, as 'the great ravelled knot'. It would take a large book just to set out the number of possible relationships between these cells, so that it is not surprising that scientists are still baffled by brain physiology. All that we can do here is to gain some familiarity with the main morphological regions of the brain and take a brief look at one example of the baffling complexity of neuron relationships.

Figure 42
Diagram to show the internal anatomy of a sheep's brain.

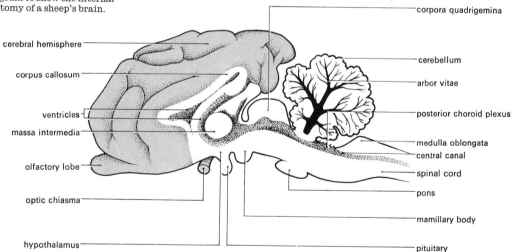

Procedure

1   Examine a specimen of mammalian brain and locate on it the main areas indicated by figure 42. Make a sketch of the specimen as seen in side view.

2   Re-examine your longitudinal slices of brain tissue and identify the main structures shown, using figure 42 to help in this. It will also help to re-assemble all the slices cut from one brain. Make an annotated sketch of the specimen brain

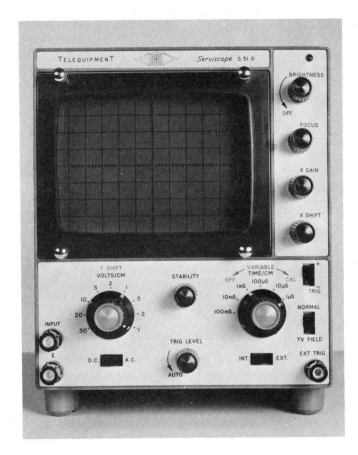

Figure 43
The cathode ray oscilloscope.
(A simple single beam oscilloscope.)

as seen in longitudinal section, cut through the mid line (*sagittal* section).

3   Examine under low and high power a permanent preparation cut from the cerebellum. Identify and sketch as many types of neuron cell bodies as you can find.

Questions

a   From your investigations what do you assume is the general distribution of neuron cell bodies and nerve fibres in the brain?

b   What form do the spaces within the brain take? Can you suggest any function for them?

c   Are the different types of neuron cell bodies located in specific regions of the cerebellum? What hypothesis can you make regarding the nature of the arrangement of the neuron cell bodies in this region of the brain?

Investigation of nerve activity

By the use of a stimulator and kymograph we can study the relationship between the stimulation of a nerve and the subsequent contraction of the muscle which it innervates. However, this gives us little information regarding the nature of the nerve impulse. To investigate this, we need to use a more sophisticated instrument, the cathode ray oscilloscope, together with a suitable pre-amplifier and associated equipment.

The cathode ray oscilloscope

This instrument operates on the same principle as the cathode ray tube of a television set. The end of the tube contains a device called an electron gun, which produces a beam of electrons. This beam is focused on the screen at the opposite end of the tube, causing a chemical coating on the inside of the tube to glow, and thus producing a spot of light. Before reaching the screen the beam passes between two sets of metal plates. These are known as the X and Y plates. One set of these (the X plates) is vertical, the second set (the Y plates) is horizontal. A potential difference between the X plates deflects the electron beam horizontally. A potential difference between the Y plates deflects the beam vertically. If the potential difference between the X plates is increased steadily, the beam traces a horizontal line of illumination across the screen.

In the oscilloscope, a steadily increasing potential difference applied to the X plates makes the beam of electrons sweep from left to right across the screen. When the beam reaches the righthand edge of the screen the potential difference drops to its initial value, the beam flies back to the lefthand edge of the screen, and the process is repeated. This produces

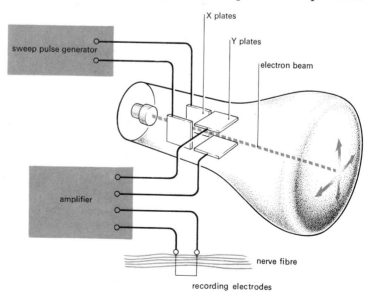

Figure 44
The cathode ray oscilloscope
arranged to measure nerve impulses.

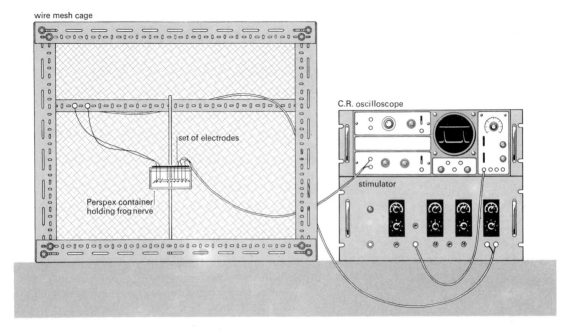

wire mesh cage

set of electrodes

Perspex container
holding frog nerve

C.R. oscilloscope

stimulator

Figure 45
The cathode ray oscilloscope and
associated equipment arranged to
investigate nerve responses in a frog.

a horizontal glowing line on the screen. If we know the frequency of these sweeps the time taken for the spot to travel
across the screen can be determined. This forms the time
base of the instrument. Use is made in neuro-physiology of
the fact that one sweep of the beam can be completed in a
matter of ten thousandths of a second. Thus it provides a
means of measuring very small intervals of time.

Any potential difference to be measured is applied to the Y
plates. The effect of this will be to deflect the spot up or down.
Thus, as the spot travels horizontally across the screen at a
steady rate determined by the time base, it will be deflected
vertically by changes in potential difference on the Y plates.
The oscilloscope can be calibrated so that any such deflec-
tions on the screen can be read off as millivolts.

Some oscilloscopes are of a double beam type in which two
lines are traced on the screen. These can be used in various
ways. For example, one beam can be used to display the
stimulus voltage being applied to a nerve, while the other
beam can record the response voltage. Permanent records of
the display on a cathode ray oscilloscope are normally made
by photographing the screen. These records can be studied
later to measure the changes in potential difference which
occurred. (Figure 46.)

Figure 46
A display on a dual beam cathode ray oscilloscope. The upper trace is due to the stimulus and the lower represents the response.

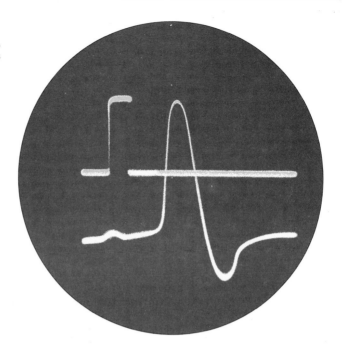

As the signal voltage from any nerve tissue is very small, it requires to be amplified a thousand or more times before being fed into the Y plates. For the same reason many precautions need to be taken to ensure that the display shows only the potential differences picked up from the nerve and not ones from other sources, e.g. from electric lights. This involves the use of special electrodes and placing the preparation in an efficiently earthed metal-mesh cage.

6.4 Use of a cathode ray oscilloscope to study the responses in a frog sciatic nerve

Although it may not be possible to investigate nerve action with a cathode ray oscilloscope in your own laboratory, you can make use of a film loop. Study the loop carefully, run it through once to observe the main sequences of the investigation, and then run it a second time and observe the details more closely. If the projector has a stop frame facility you may find this useful to study details.

Procedure

1   Dissect the frog and carefully remove the sciatic nerve.
2   Prepare the electronic apparatus consisting of a double beam oscilloscope, amplifiers, and stimulator. Turn up the stimulus voltage and check the display on the tube, then reduce the voltage to zero again.
3   Place the preparation on a set of electrodes in a Perspex bath. (In the loop the two lefthand electrodes are stimulating electrodes and the two righthand ones recording elec-

trodes. The earth electrode was positioned next to these. The other three electrodes were not in use in this investigation.) The bath should contain a little Ringer's solution. A cotton wick may be attached to a middle electrode and allowed to dip into the solution. Replace the lid of the bath.

4   Connect up the electrodes to the oscilloscope. Increase the voltage control to the stimulating electrodes from zero and note the response of both beams on the display, and then decrease the voltage.

5   Operate the delay control on the stimulator to allow two stimuli to be delivered with a very short time interval between them. Gradually reduce the time period to zero and then increase it again.

Questions

a   What advantage might there be in using glass tools to separate the nerve from the muscle tissue?

b   What was shown on the display tube when the stimulus voltage was increased?

c   What was the purpose of the Ringer's solution?

d   What was the function of the cotton wick attached to one of the middle electrodes?

e   When the stimulus voltage was increased, the nerve to the stimulating electrodes showed a compound action potential. Why is it called a compound action potential?

f   Can you explain what happened after the nerve had shown a clear compound action potential, when you increased the stimulus voltage still further?

g   The delay control on the stimulator allows two stimuli to be delivered with a very short time interval between them. What was the effect of these two stimuli on the nerve response?

h   When the time period, i.e. the distance seen on the cathode ray oscilloscope between the two stimuli was gradually reduced, what effect did this have on both responses? Can you suggest an explanation for these effects?

6.5 Responses in the locust's central nervous system

The cathode ray oscilloscope can also be used to investigate the activity of the central nervous system and the effect on this activity of the stimulation of various sense receptors. An insect, such as a locust, provides a convenient organism for this work. Again a film loop will be helpful when it is not possible to undertake the actual preparation.

Procedure

1   Dissect a locust to expose the ventral nerve cord. Place the preparation inside a wire mesh cage.

2   Place two stimulating electrodes, two recording electrodes, and an earth in contact with the nerve cord, and connect these up with the cathode ray oscilloscope and stimulator.

3   Without stimulating the nerve cord, observe on the display any electrical activity from the nerve cord.

4   Stimulate an anal cercus several times with an artist's paint brush and observe the effect on the display.

5   Now stimulate the nerve cord via the stimulating electrodes. Gradually increase the stimulus voltage and then decrease it.

Questions

a   Why do you think the investigation was made within a wire mesh cage?

b   When you did not stimulate the nerve cord what could be observed on the cathode ray oscilloscope?

c   How can you explain the observed effect?

d   What effect was seen when you stimulated an anal cercus?

e   When you stimulated the nerve cord via the stimulating electrodes what was the effect on the response of a gradually increasing stimulus voltage?

f   In both locust and frog we were observing compound action potentials. Assume that the cathode ray oscilloscope has been set up in a similar fashion in each case, i.e. that the same time base was used, and that each preparation nevertheless showed a different pattern. Can you suggest an explanation for these differences?

g   What was the effect on the response of reducing the stimulus voltage to the nerve cord? How could you explain this effect?

Bibliography

Baker, P.F. (1966) 'The nerve axon.' *Scientific American* Offprint No. 1038. (Experimental studies using a single nerve fibre from the squid.)

Brewer, C.V. (1961) *The organization of the central nervous system*. Heinemann Educational Books. (Good general account.)

Clegg, A.G. and Clegg, P.C. (1962) *Biology of the mammal*. Heinemann Medical Books. (Account of nerve impulse and the brain.)

Freeman, W. H. and Bracegirdle, B. (1966) *An atlas of histology*. Heinemann Educational Books. (Full histological study of nerve tissues.)

Gray, G.W. (1948) 'The great ravelled knot.' *Scientific American* Offprint No. 13. (Straightfoward account of the structure and functioning of the brain as known in the late 1940s.)

Katz, B. (1952) 'The nerve impulse.' *Scientific American* Offprint No. 20. (Clear account of the nerve impulse as known in the early 1950s.)

Marshall, P. T. and Hughes, G. M. (1965) *Vertebrate physiology*. Cambridge University Press. (Chapter on nervous coordination.)

Schmidt-Nielsen, K. (1960) *Animal physiology*. Prentice-Hall. (Chapter on 'Integration'.)

Synopsis

1 The relationship between strength and response of a stimulus can be explored in single axon fibres.

2 A motor axon forms an intimate relationship with an individual muscle fibre called a motor end plate.

3 Action potentials pass down the axon to motor end plates and generate an action potential in the muscle fibres.

4 A nerve conducts impulses at a measureable speed.

Extension work II

Neurophysiology

Summary of practical work

section   topic

E2.1   Effect of a varying stimulus strength applied to the motor axons supplying the muscles of a locust leg

E2.2   Response of skeletal muscle

E2.3   Velocity of a nerve impulse

In our investigations into the nature of muscle twitch in Chapter 5 we stimulated a complete nerve composed of several hundred axons. As a result we could only see the effect of the stimulus on the nerve as a whole. To investigate the effect of stimuli on a nerve axon we can obtain a more suitable preparation from a locust. It is known that the nerve supply to the extensor muscle of the jumping leg of the locust contains few nerve axons. In *Schistocerca* sp. only three motor axons supply the muscle. There are two large excitatory axons, one being a 'fast' fibre, responding rapidly to a single adequate stimulus, and the other a 'slow' fibre. The slow fibre only comes into operation when the number of stimuli per second is over 15. The function of the third small axon is uncertain but it may be an inhibitory fibre.

E2.1 Effect of a varying stimulus strength applied to the motor axons supplying the muscles of a locust leg

If the motor axons to the leg muscles of the locust are stimulated, both the extensor and flexor muscles will contract. However we can nullify the effect of the contractions of the flexor muscle by cutting the tendon connecting the muscle to the exoskeleton. In these circumstances we shall only see the effect of the extensor muscle and can record it on a rotating drum.

Procedure

1 Anaesthetize a locust, hold it by the wings, and remove the head, abdomen, and first two pairs of legs. Pull out the gut remaining within the thorax.

2 Place the thorax, ventral side uppermost, on a block of cork, and pin it in position through the anterior portion of a sternal and tergal plate. Remove the wings.

3 Secure both femora to the cork by cross pins, as shown in figure 47.

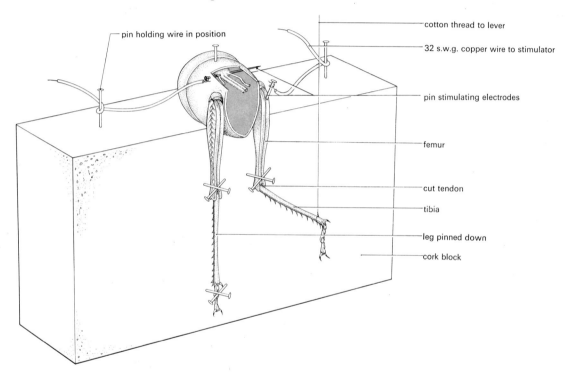

Figure 47
The femora of the locust secured to the cork.

Place the preparation under a low power binocular micro-scope and identify the tendon of the flexor muscle at the femorotibial joint as shown in figure 48.

Sever the tendon, using the tip of a sharp scalpel.

4 Make a superficial lateral incision on either side of the meta-thoracic sternal plate. Lift it up and cut across transversely to remove it. Be very careful not to remove any underlying tissue. Remove any other portions of the integument as necessary, until the metathoracic ganglion is exposed. If necessary, moisten the preparation with a drop or two of invertebrate Ringer's solution from time to time.

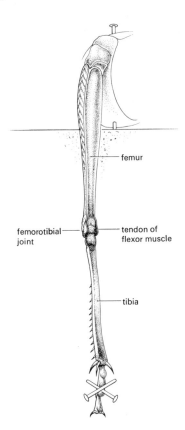

Figure 48
The leg joint of the prepared locust.

femur

femorotibial
joint

tendon of
flexor muscle

tibia

5   Connect a fine entomological pin to the stimulator, using
    about 32 SWG covered copper wire. Pierce the thorax later-
    ally with this pin so that it passes through the ganglion.
    Using a similar copper wire connect the stimulator to ano-
    ther pin. Insert this pin obliquely into the side of the femur.
    Ideally, this pin should be as near to the nerve supplying the
    leg as possible. It may be advisable to pin the stimulator
    leads to the cork block.
6   Tie a length of cotton thread to the tibia of the leg with the
    severed tendon. Arrange the apparatus so that the other end
    of the thread can be attached to the end of an ink writing
    lever. Adjust the lever so that the tibia is horizontal.
7   Set the voltage control to give about 1·0 volts and deliver a
    single shock to the preparation. If you do not think that the
    leg has given a full response, increase the voltage until it
    makes such a response. Arrange the recording drum at a
    suitable height and move the apparatus until the ink pen is
    in a position to record on the drum. Set the stimulator to
    deliver 1 shock per second, and the recording drum speed to
    about 2·5 mm per second. Switch on the apparatus.

8 Record several responses to the stimuli, then slowly reduce the voltage step by step until a point is reached when the leg no longer twitches. Now increase the voltage step by step to produce 10 to 15 responses, then reduce it slowly again.

9 Remove the paper and note the strength of each stimulus beneath the appropriate response.

Questions

a Observe the walking and jumping movements of a normal locust. How will contractions in the extensor muscle of the hind limb affect its movement?

b What is the effect on the response of stimulating the preparation with a range of voltages?

c What explanation can you suggest for the results?

d What is the effect on the response if the preparation is stimulated with stimuli of adequate voltage but increasing in frequency of 1, 2, 5, 10, and 20 stimuli per second? (Give the preparation time to recover between each burst.)

Physiology of skeletal muscle

The relationship between a skeletal muscle and its motor nerve supply is a complex one. Any motor neuron in the central nervous system sends out a single axon fibre. A nerve such as the sciatic is a bundle of axons. Each motor axon divides, as it enters a muscle, into 5 to 100 branches, each of which ends at a neuromuscular junction on a single muscle fibre as illustrated in figure 49.

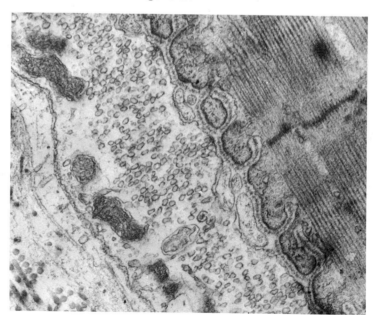

Figure 49
Electron micrograph of neuromuscular junction of a frog.
From Birks, R., Huxley, H. E., and Katz, B. (1960) 'The fine structure of the neuromuscular function of the frog'. J. Physiol. Lond., **150**, *No. 1., pp. 134–144.*

The ultimate end of the branch of the axon is called a motor end-plate. The whole assemblage consisting of the neuron cell body, axon, its branches, and all muscle fibres connected to them responds as a single motor unit.

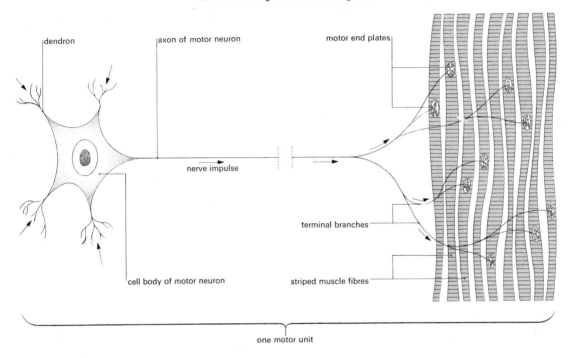

dendron   axon of motor neuron   motor end plates

nerve impulse

cell body of motor neuron   terminal branches

striped muscle fibres

one motor unit

Figure 50
Diagram of a motor unit.

When the motor neuron fires, the action potential passes down the axon and its branches will, in turn, generate muscle action potentials that pass along the muscle fibres. Muscle action potentials cause the contraction of muscles. Thus, all the muscle fibres supplied with branches from one axon will contract when an impulse passes down that axon.

E2.2 Response of skeletal muscle

We have already seen (5.4) that the strength of the stimulus applied to a complete nerve such as the sciatic nerve is important, for there is a minimum level of the stimulus, below which no effect will be produced. The following investigation is to record the amount of muscle contraction when we raise the strength of stimulus slowly from the minimum value.

Procedure   1   Set up a frog sciatic nerve/gastrocnemius muscle preparation as before (see 5.3).

2 Arrange the apparatus to record on a stationary drum.
3 Record contractions of the muscle at stimuli of increasing strengths, starting from the lowest voltage that just gives a response.
4 Turn the drum by hand about 1 cm.
5 Increase the voltage slightly and repeat the recording. Continue this until an increase in the stimulus voltage gives no further increase in the amount of the contraction.
6 Remove the paper from the drum and below each stimulus record the magnitude of the stimulus.

Questions

a Between the limits of a stimulus which just gives a response and one that produces a maximum reaction, what is the relationship between the size of stimulus and that of the response?
b How could you explain the variation in response in terms of the operation of motor units?
c How could you test your explanation?

E2.3 Velocity of a nerve impulse

Muscles contract rapidly when stimulated by nerve impulses. The speed at which a muscle responds is due to the high velocity of the motor nerve impulses from the central nervous system, and the relatively short distances that they have to travel. The velocity of nerve impulses can be investigated, using a sciatic nerve/gastrocnemius preparation. However, as the length of nerve which can be used is short and the maximum speed of a kymograph is not very high, it will only just be possible to measure such a small time interval. The procedure is simular to the classical experiment performed by Helmholtz in 1850 when he first discovered that it was possible to measure nerve impulse velocities.

Procedure

1 Make a standard frog sciatic nerve/gastrocnemius muscle preparation. Take care to dissect out as long a portion of sciatic nerve as possible. Place the preparation in a bath of Ringer's solution at 5 to 10° C.
2 Set up the preparation for recording, but use two sets of stimulating electrodes. Place one set under the nerve, as near to the muscle as possible, and the second set as far away as possible.
3 Apply stimuli of increasing strength to the preparation through the electrodes nearest to the muscle until you obtain a maximum response. Leave the setting on the stimulator at this level.
4 Complete the circuit for one drum triggered stimulus to be delivered for each rotation of the drum. Set the kymograph to rotate at its maximum speed.

5   Start the kymograph, allow it to pick up sufficient speed, and then arrange a drum triggered stimulus to be delivered through the electrodes nearest to the muscle. Stop the kymograph.

6   Without moving the preparation, drum, or lever, disconnect the electrodes nearest to the muscle and connect the set further away. Then repeat the procedure in (5) above.

7   Remove the paper and measure the distance between the cathodes (negative) of the two electrodes.

8   It may be possible by raising or lowering the drum to obtain two records on one sheet of paper. If not, use a fresh sheet and repeat the operation.

Questions

a   The two records should appear with a small time interval between them. Measure this interval. Knowing the distance between the electrodes, calculate what velocity of conduction this represents.

b   What was the purpose of using cold Ringer's solution for bathing the preparation?

c   If you have reasonably consistent results, you might investigate the effect on the velocity of changing the temperature of the Ringer's solution bath by 10° C. What effect does this have?

d   What modification of the kymograph would give a more accurate result?

e   What instrument would provide an even more accurate way of measuring nerve impulse velocities?

Review
Question

Summarize the present ideas on the nature of the nerve impulse.

Bibliography

Dawson, H. (1964) *A textbook of general physiology*. Churchill. (Detailed account of nerve action.)
Lippold, O.C.J. and Winton, F.R. (1968) *Human physiology*. 6th edition. Churchill. (Chapter 17, 'Nerve'.)

Chapter 7

Social behaviour

Summary of practical work

Co-ordination in animals

So far we have been considering some of the ways in which cells, organs, or individual organisms are able to co-ordinate their activities in response to changes in their environment.

Individual animals often form groups or aggregations and the group or individuals may be able to respond to changes in such a way that all benefit. Human societies are excellent examples of this type of behaviour.

The bonds keeping animals together are of a number of different kinds. Sometimes animals are attracted together in response to a common environmental stimulus, as when woodlice congregate under a stone or the grubs of *Apanteles* live collectively inside a larva of *Pieris brassicae* and may show a behaviour pattern specific to their species.

There is some evidence that, in addition to such factors as humidity and light, scent may play a part in causing some species of woodlice to remain together. This is certainly true of the honey bee, where a secretion of the queen (queen substance) plays a vital part in maintaining the hive as a social entity. (See figures 51 and 52.) The workers also produce characteristic scents which serve the same purpose. Chemical secretions produced by animals for mutual recognition. thus influencing behaviour, are sometimes known as *pheromones*. Among the larger animals, group behaviour becomes more obvious and commonplace. Thus we find sheep existing in flocks, lions in prides, geese in gaggles, and wolves in packs. In every instance the social bonds involved are only incompletely known but derive from the peculiar attributes of the animals themselves.

Figure 51
A worker honeybee disseminating scent.
Photo, C. G. Butler, F.R.P.S.

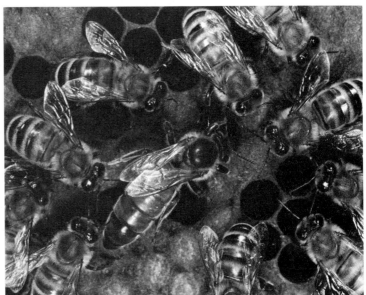

Figure 52
A 'court' of worker honeybees around their queen.
Photo, C. G. Butler, F.R.P.S.

Advantages of a social existence

The fact that many animals exhibit some degree of social organization suggests that there must be advantages to be gained from this mode of life. The principal of these can be summarized as follows.

1 Efficient food collecting – This is particularly well illustrated by the honeybee where the workers are able to communicate by means of their peculiar 'dance' the magnitude of a source of food and its direction from the hive.

2 Defence against predators – Many animals adopt a particular type of behaviour to signify the approach of an enemy, such as the characteristic 'kew kew' call of the house sparrow on the approach of a marauding cat.

3 More efficient breeding – Some degree of social behaviour will have the obvious advantage of increasing the probability that a male will find a mate. The reproductive behaviour of many animals shows a period before mating takes place when the 'engagement' of the pair of animals is obvious. This close association is known as a pair bond and has particular advantages for animals living in groups. It is shown in many species and is, of course, a well known feature of human courtship. Other effects on breeding may be more subtle. For example, in the black-headed gull, which normally breeds in colonies, it has been shown that broods reared outside the colony area are markedly less successful than those reared within it.

4 Quicker and more efficient learning – Probably all animals are capable of some degree of learning in the sense that later behaviour is affected by earlier experience. Within a group, the possibility of learning from the diversity of experience available is greatly enhanced. In some groups of social animals, particularly the primates (lemurs, monkeys, and man) this plays a vital part in development. Experiments have shown that some animals reared in isolation prove to be incapable of playing a normal role in the community later on. The conveying of information within a human society is much more efficient than in any other species owing to the highly developed means of group communications – written and spoken word and the means to transmit these rapidly.

Figure 53
Larvae of the large white butterfly.
Photo, S. Beaufoy, F.R.P.S., F.C.S.E.

5 Other factors–There are probably a number of other advantages to be gained from a social existence which are not
easily classified under a single heading. For instance, the
larvae of many insects such as butterflies often live in
aggregations, particularly during their early phases. This is
true of species which possess warning colours indicating
noxious taste such as the large white butterfly, *Pieris
brassicae*, and those with irritant protective devices such as
the spines of the caterpillar of the small tortoiseshell
butterfly, *Aglais urticae*. Not only does such behaviour reduce
the possibility of predation, but the temperature of the
group is thereby raised, thus speeding up development.
(Figures 53 and 54.)

Figure 54
Young, small larvae of the
tortoiseshell butterfly in a web.
Photo, S. Beaufoy, F.R.P.S., F.C.S.E.

Social order

Judged superficially, animals which live a gregarious
existence appear to exhibit little evidence of any sort of
social organization. But in the 1920s a Norwegian scientist,
Schjelderup-Ebbe, made a fundamental discovery in animal
behaviour. He carefully studied a flock of hens and, as a
result of these observations, was able to show that there was
a distinct pattern to the interactions going on within the
group. For instance, in an established flock, when two
particular birds met, one would adopt an aggressive attitude
towards the other and the second would then respond, not by
retaliation, but by appeasing and retreating. The aggressive

bird would drive the other away by pecking at it or by merely threatening to do so. This type of situation was found to exist throughout the flock, and what was apparently just a random group of birds, turned out to be a highly structured society. Such a social hierarchy was called by Schjelderup-Ebbe a *peck order*. Since that time workers in the field of animal behaviour have discovered this type of social organization among other animals. The pattern may vary in different species as does the intensity with which social predominance is expressed. The type of society found in hens also occurs among geese and swans and is sometimes known as a 'closed' society. By contrast, the situation among African big game and in flocks of starlings during autumn differs in that dominance does not seem to occur. Such societies are therefore referred to as 'open'.

7.1 Peck order in hens

As you can guess, investigating a flock of hens in order to determine the peck order is time-consuming. However, the problem can be greatly simplified if we use a film of the different pairs of birds in action. In it, each bird has been marked with a different colour and the various sequences show what happens when the birds are caged in pairs in every possible colour combination. Before starting the film, equip yourself with a pencil and paper for recording the observations you will need to make.

Procedure

1   Run the film through a first time without making any records, just to see the general trend of events. Is there anything in common between the behaviour exhibited by each pair of birds?

2   Start the film a second time and record the result of each pairing (which bird predominates and which one appeases). Note the essentially ritual nature of each encounter; the appeasing bird is not seriously hurt.

3   When you have completed your records, work out the peck order of the hens as indicated by their colours.

Questions

a   In what ways does the predominating bird exert its dominance?

b   How is appeasement signalled by the other bird?

c   What do you think may be the function of an established peck order in hens?

d   Some types of human society would appear to have a rather rigid system analogous to the peck order of hens. The army would, perhaps, be a good example of this. Do you think that a less rigid and obvious system might exist in your own school? If so, what signals are used to indicate dominance and appeasement?

e   What other examples of 'peck orders' can you think of in human society? How do they differ from that exhibited amongst hens?

Co-ordination and reproduction: courtship

One of the most fundamental aspects of co-operation among vertebrates is that involved in reproduction. As a preliminary to mating some form of courtship normally takes place. Courtship has several important biological functions:

1 It allows the two sexes to recognize each other.
2 It ensures that the two sexes are ready to mate.
3 It enables closely related species to distinguish each other.
4 It may serve as a stimulus to operate internal control mechanisms concerned with reproductive processes, such as ovulation.

7.2 Courtship of the stickleback, *Gasterosteus aculeatus*

As in all forms of co-operation, in courtship communication is involved between the participants. The classical work in analysing the courtship behaviour of an animal was carried out by Professor Tinbergen on the common three-spined stickleback, *Gasterosteus aculeatus*, which can be found in nearly every pond.

The investigations were made in two main ways. First, courtship of the animals was observed in large aquarium tanks under as natural conditions as possible. By timing the events and studying these closely, the investigators discerned the pattern of courtship. This raised a number of problems, particularly the need to discover exactly what were the stimuli to which the animals responded at different stages in the behaviour sequence. Second, to investigate this aspect, the investigators used a more experimental technique, whereby they presented various models to the fish and stimulated them in other ways. From this work a fairly complete picture emerged both of the events themselves and the mechanisms behind them.

One of the difficulties in studying behaviour is that animals will seldom behave to order. With patience you can observe stickleback courtship in an aquarium during early summer. Failing that, make use of a film.

Procedure

1   Set up an aquarium with suitable gravel, water plants, etc., and place in it one male fish in breeding condition.
2   Add a female fish, in breeding condition, immediately after you have added the male to the tank and observe and record any interactions which take place over the next 10 to 15 minutes. In your record include a note of the time periods for activity. If the female is attacked vigorously during this period, remove her; in any event remove her after the 15 minutes.
3   Observe and record the activities of the male from time to time over the next few days. Normally male fish construct a nest among plant debris. Ensure that there are suitable materials for this.
4   When a nest has been successfully completed, re-introduce the female fish and carefully record all activities and the time that they take. Again, if the male shows any danger-ously aggressive behaviour towards the female, remove her.
5   If courtship is successful and leads to the laying of eggs, continue your observations of the male fish during the time the eggs are developing and after they hatch.
6   If you first study the behaviour of sticklebacks by means of a film, try at some later stage to compare the conclusions you have drawn with those made from observations on actual animals.

Questions

a   What physical events must precede a successful courtship?
b   What processes were involved in building the nest? What discrimination did the male show during this process? If a pencil is moved slowly through the water near the nest what effect does this have on the male?
c   Describe the movement of a breeding male towards a recep-tive female. By what means do you think a male may identify a female? If possible, try to test your hypothesis on a male fish.
d   Describe the separate processes which occur from the time that the male swims towards the receptive female until the eggs have been fertilized. What stimulus is used to elicit egg laying? How could you test your hypothesis?
e   Compare the effects of adding another receptive female im-mediately after a successful spawning with the effects of adding one 24 hours later.
f   Summarize the responses of courtship in the form of a chart.

Experimental study of courtship

Two general conclusions have emerged from the experi-mental study of courtship in animals.
1 Courtship often follows a rigid sequence of events. Each event is brought into action by the successful conclusion of the preceding one, and all of them involve an interaction between the male and the female forming a pattern, illus-trated in figure 55 for the stickleback. This complex chain

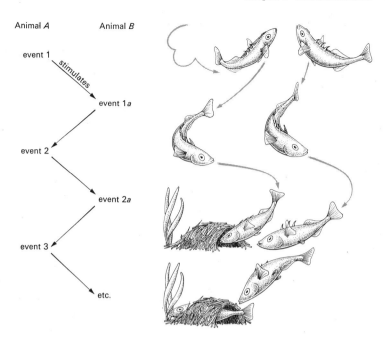

Figure 55
The sequence of courtship of the three-spined stickleback.
Partly based on Tinbergen, N. (1951) The study of instinct. *Oxford University Press.*

Figure 56
The feeding response of the *Dytiscus* beetle displayed within an area of chemical stimulus set up by diffusion from a cloth bag containing flesh. What control would be suitable for this experiment?
After Tinbergen, N. (1951) The study of instinct, *Oxford University Press.*

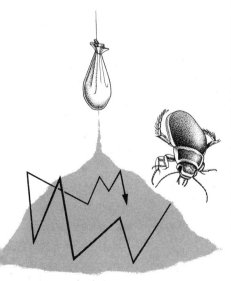

of events leads to gamete release and fertilization of the eggs. In the stickleback, at least, the pattern is immutable and the failure of any one link in the chain results in the cessation of courtship. In other animals the behaviour patterns may not be as rigidly set.

2 It was found that the animals responded to one particular stimulus in the environment and not to the whole range of stimuli available to them through their sense organs. The specific stimulus which produces a particular pattern of behaviour is called a *sign stimulus*. This type of behaviour has been observed in many other animals, for example, the carnivorous water beetle, *Dytiscus marginalis*, which, although it has well developed compound eyes, does not use them when hunting prey. Instead, the feeding responses of the beetle are released by chemical and tactile stimuli only. Thus in the presence of a watery meat extract a hungry beetle will 'hunt' and 'capture' every solid object that it touches (figure 56). The fact that an animal may make an automatic response to only one stimulus in its environment means that it can be 'fooled' in similar ways. It may possess sense organs perfectly capable of discrimination, but it can only respond to the sign stimulus. Song birds finding a cuckoo in their nest will feed it, although it looks very different from their own young. Its beak gapes widely, and by showing a brightly coloured mouth and throat it stimulates the foster parents to feeding activity. In contrast, they will ignore one of their own young pushed onto the edges of the nest, as if it no longer existed.

Communication

From the examples of co-ordination among animals examined so far, it will be seen that much depends upon the ability of the individuals to communicate information to one another. In establishing a peck order the predominant hens asserted their supremacy by particular behaviour patterns. However, communication is frequently of a more subtle kind and the response to sign stimuli is inborn, thus rendering learning unnecessary. For example, in the herring gull the alarm call made by the parents releases a hiding response in the chicks, but the call cannot convey the direction in which the danger lies. Tinbergen relates that one day, when he was taking photographs from a hide in a gull colony, he made a careless movement. The parents in a nearby nest immediately gave the alarm call. The young ran straight into the hide, and crouched at his feet.

Figure 57
A black-headed gull rolling an oversized dummy egg into a nest in preference to a normally sized dummy.
Photo, Professor N. Tinbergen.

On the other hand, behaviour studies have shown that a stimulus which appears to be of quite a simple kind can, nonetheless, convey a great deal of information. In some instances it has been possible to break this down into its components and to set up models to test their individual

effects. Thus the process of incubation in birds is initiated and maintained by a number of influences, some of which are still unknown.

One of these is, undoubtedly, the presence of an egg in the nest and has been fully demonstrated in the herring gull. Evidently, it does not have to be a natural egg, for, provided the object in the nest bears a superficial resemblance to the real thing, the necessary information is conveyed to the female whose powers of discrimination are strictly limited. Thus, the addition of a 'super' egg to the nest which already contained a normal one resulted in the bird attempting to incubate the fake in preference to its own, although it was so large that the animal could barely sit upon it.

Chemical signals

A considerable amount of information can be communicated by chemical substances. The world of a dog appears to be largely circumscribed by a variety of smells and the dog's process of smell discrimination is correspondingly high. Chemicals such as those released in urine are used to mark territories and home ranges, to proclaim a warning signal to intruders, and to attract a sexual partner. In this respect, the attractiveness to male dogs of vertical objects, such as lamp posts and trees is proverbial, although the function of this behaviour has not yet been definitely established.

Visual signals

Visual communication is used to a considerable extent by many animals, in particular by the higher vertebrates, to indicate their 'emotional' state. Figure 58 shows the varied facial expressions and postures adopted by cats which convey a wide range of information.

Auditory signals

Although a wide variety of animals use auditory communication, the range of sounds produced is only large in birds and primates. Analysis of the song of the robin has shown that a particular bird may perform as many as 57 different variants and it may well be that individuals can be recognized by others through their peculiar range. It has been shown, for instance, that male chaffinches can distinguish between established rivals and newcomers to the area. The range of different calls has been recorded for certain species of primates: about 13 for gibbons and 15–20 for howler monkeys. This is in striking contrast to the diversity of human utterances. When evaluating the difference, we must bear in mind that while the capacity to produce auditory signals in animals is largely inborn, human language is acquired anew at each generation as a result of imitation and experience. Moreover, the diversity of sounds that man is capable of producing has resulted in a highly complex and descriptive system of auditory communication.

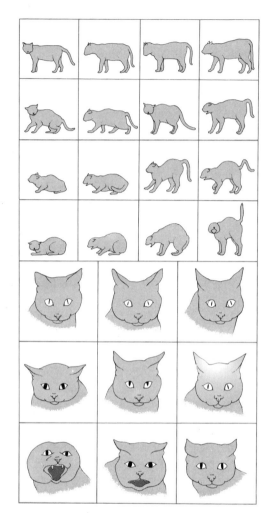

Figure 58
Communication in cats by means of body and facial expressions. In each section, aggressiveness increases from left to right and 'fear' increases downwards. This provides a graded series of signals.
From Leyhausen, P. (1956) 'Das Verhalten der Katzen (Felidae)', Handbuch der Zoologie, **10,** *(21) 23, Abb. 23, Walter de Gruyter.*

7.3 Co-ordination and family life

In mammals, most birds, and some other animals, co-ordinating behaviour has been extended to caring for the young, and this has obvious survival value for the species. This often involves a complex interaction between the parents, and the parents and their offspring. This type of behaviour has been extensively studied in gulls and terns.

Procedure

Examine a film showing some of the activities found in the nest area of common terns *(Sterna hirundo)* in the breeding season. Use the film as you did the others, running through the first time to observe the main trends of the behaviour of the terns, and then a second time to observe the details more closely.

Questions

a Does the film provide any evidence of territorial behaviour in the nest area? If so, what advantages do you think it affords the birds?

b How do the parents respond to an intrusion near to their nest?

c What functions would you ascribe to the removal of the egg shells?

d How does a chick respond to the stimulus of hunger? How could your hypothesis be investigated? What is the effect of the chick's behaviour on the parent?

e Can you think of any human behaviour which appears analogous to that of the tern? Do you think that any sign stimuli may be operating in the relations between human parents and children?

Co-ordination in man

Man is probably the best example of a social animal. Our modern industrial society has certainly the most complex form of any animal group. Such an organization can only function with a built-in system of social controls. This provides a compromise between the needs of the individual and those of the society. Social controls include various aspects of government, the legal system, education, social welfare, etc. Many aspects of man's social behaviour, such as the forms of social greeting, are rituals and provide an accepted norm of behaviour for communication within the society.

Perhaps the main reason why such a society is viable and man's various activities can be co-ordinated is that man possesses a greater ability to learn than any other animal. In fact, it is this feature of man as an animal which most clearly separates him from all other animals, even from his nearest biological relatives, the apes.

Learning is not an easy word to define. In broad terms it could be said to be an adaptive change in behaviour as the result of experience. Learning in this sense is found in all but the simplest forms of animals. The process of learning is itself an extremely complex subject and is, of course, one of the main areas of research among psychologists and educationalists. Although a detailed study of learning is beyond the scope of this chapter, there are some questions we can investigate regarding the way in which learning processes may occur.

Many learning situations are complex and the kind of learning involves more than one act. The order in which the separate acts occur may be dictated by the learning situation. Learning to drive a car is one of these situations. It involves learning to use the steering wheel with one hand and the gear lever with the other, while the feet are involved with the clutch, brake, and throttle controls. Psychologists have devised simple laboratory experiments to investigate this type of learning. One of these is mirror drawing.

7.4 Mirror drawing

When we draw, we use a form of skill called *sensorimotor*. That is a skill where muscular movement is important but is under sensory control. The muscle movement is not simply a pattern of movement but is altered by feed-back through the sensory system.

In mirror drawing our normal co-ordination between hand and eye is useless and we have to learn a new pattern.

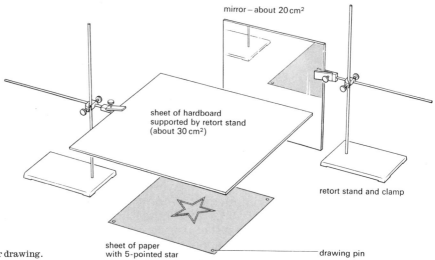

mirror – about 20 cm²

sheet of hardboard
supported by retort stand
(about 30 cm²)

retort stand and clamp

Figure 59
The apparatus for mirror drawing.

sheet of paper
with 5-pointed star

drawing pin

Procedure

1   You should work in pairs for this investigation. One partner should act as the experimental subject while the other records, and then they should exchange roles. First fix a suitable mirror in an upright position on the bench, about 18 inches away from the subject.

2   Arrange a sheet of hardboard or other suitable material in a retort stand as shown in figure 59. The sheet should be high enough to enable the subject to write with a pencil beneath it, but not so high as to block the mirror.

3   Place the 5-pointed star figure, which has a double outline, beneath the sheet. At a given signal, the subject should trace a pencil line around the star and in between the two lines. He should do this as quickly as possible, trying to avoid crossing either line with his pencil. He must not touch the bench with any part of his body.

4   Repeat this for 10 to 12 trials, using a fresh star figure each time. Pin in the same place and start tracing from the same point.

5   Graph the results as follows:
   1 Time of each trial against number of the trial.
   2 Number of errors, i.e. the number of times the pencil line crossed one of the printed lines of the star figure.

Questions

a   What does the graph indicate about the speed at which you can learn this type of skill?

b   Assess the rate of learning for the other members of your class. How much variation do you get?

c   Does the skill you have gained on this particular drawing make it easier for you to produce a different mirror drawing? How would you know whether it did or not?

7.5 The influence of experience

Learning, as we have seen, is an adaptive change in behaviour as the result of experience, and we all hope to learn from our experiences. We can investigate the effect, upon a subsequent action, of knowing the result of a previous one, by simple experiments.

Procedure A

1   Again work in pairs, using one partner as the experimental subject. For the first investigation you require a number of blank sheets of paper. 10 cm × 20 cm is a convenient size.

2   The subject should have a pile of paper in front of him. He should take the first sheet and draw freehand a straight line as near 7·5 cm long as he can. The piece of paper should then be turned face down on the bench and a line drawn on the second sheet as before. Continue this for 20 trials.

3   The 21st sheet of paper should be handed to the experimenter who measures it to the nearest tenth of an inch and then tells the subject the result. Only then should the subject draw a line on the next sheet. Again continue like this until a second series of 20 sheets has been obtained.

4   Measure and record the lines in the first series and the second series. Draw a graph of the length of line against the number of the trial series and include both series on the same graph.

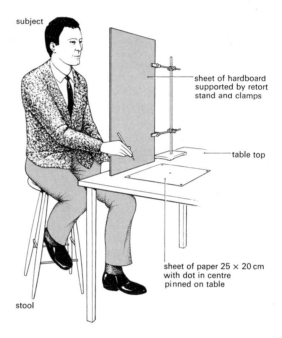

subject

sheet of hardboard supported by retort stand and clamps

table top

sheet of paper 25 × 20 cm with dot in centre pinned on table

stool

Figure 60
Dot learning test.

Procedure B

1   Set up the experiment shown in figure 60.

2   The subject should be sitting comfortably and he should be easily able to mark the sheet of paper around the dot, while unable to see it.
The subject should endeavour to mark the paper as near to the dot as possible. His arm should not touch the bench top and he should withdraw it after each attempt. At this point, each time, the experimenter should measure the distance between the central dot and the pencil mark and then the subject should make another mark. Continue in this fashion for 20 trials.

3   The series should then be repeated, but this time the subject should be informed of the previous trial, e.g. '5 cm to your left and away from you'.

4   Again graph the results of the two series of trials on the same graph.

Questions

a   Did these investigations indicate that knowledge of the results enhanced a subsequent performance?

b   If so, how do you think this information enabled the subject to perform better?

c   What would be the effect on performance if you ran a series of about 50 trials without a break with knowledge of the results?

d   On what occasions are the results of this investigation used in everyday life?

The books marked with an asterisk contain good accounts of most of the aspects of behaviour discussed in this chapter. You should read the relevant sections of at least one of these.

Bibliography

Andrew, R.J. (1965) 'The origins of facial expressions.' *Scientific American* Offprint No. 627. (Use of expression as a means of communication.)

*Carthy, J.D. (1965) *Animal behaviour*. Aldus books. (Well illustrated book.)

*Carthy, J.D. (1966) *The study of behaviour*. Edward Arnold.

Collias, N.E. (1964) *Animal language*, B.S.C.S. Pamphlet No. 20. D.C. Heath. (Communication in animals.)

Lorenz, K. (1952) *King Solomon's ring*. Methuen. (Entertaining account of some aspects of social behaviour and interactions between animals and man.)

Skinner, B.F. (1951) 'How to teach animals.' *Scientific American* Offprint No. 423 (The teaching of animals, based on psychological research.)

Tinbergen, N. (1953) *Social behaviour in animals*. Methuen.

*Tinbergen, N. (1966) *Animal behaviour*. Time-Life International. (Very well illustrated account.)

Washburn, S.L. and DeVore, I. (1961) 'The social life of baboons.' *Scientific American* Offprint No. 471. (Interesting account of a non-human primate society.)

Synopsis

1 Movement in animals is influenced by a large number of different stimuli.

2 Carefully designed experiments are needed to investigate the effect of any one stimulus on the orientation of an animal.

3 Courtship behaviour can be conveniently studied in cage birds and extended to common wild species.

Extension work III

Further study of behaviour

Summary of practical work

section   topic

E3.1   Investigation of behaviour in Protozoa *(Spirostomum)*

E3.2   Behaviour of molluscs

E3.3   Courtship behaviour in the zebra finch (*Taeniopygia castanolis)*

Studying behaviour

As you will probably have discovered already, studying the behaviour of animals is beset with difficulties. For example, the animals may not behave as expected at the time they are observed. Within a population of a single species there may be a wide variation in responses to apparently identical circumstances. Animals may differ in their responses throughout a twenty-four hour period. Moreover, they may show varying behaviour patterns at different stages of development.

All this means that it is difficult to lay down rigid experimental procedures for investigations of behaviour. Instead, some broad outlines of investigation are suggested which will necessarily have to be adapted for particular circumstances.

Orientation behaviour

In Chapter 1 you studied the orientation behaviour of *Tribolium* beetles in a humidity gradient and in Chapter 4 the orientation of *Calliphora* larvae in a light beam.

Light and humidity are two important factors which affect the orientation of many animals. Other factors such as gravity, chemical factors, temperature gradients, and sounds, may also influence an animal's orientation movements. Many of the invertebrate animals that you have studied, such as woodlice, *Tribolium* and *Calliphora* larvae, *Sitophilus* and Protozoa, can be utilized and suitable experiments devised to investigate their responses to specific stimuli.

In devising these experiments great care is required in experimental design to ensure an adequately controlled situation, so that you can be reasonably sure of the stimulus to which the animal is responding. Furthermore, it is important to relate behaviour you have observed in the laboratory to that occurring in the natural habitat.

The procedures outlined below are suggestions for the sort of work you might attempt. Obviously this will be determined by the availability of organisms and equipment, and the amount of time you have available.

It is far better to carry out an adequate investigation of one organism responding to a single stimulus than to attempt to experiment with several organisms and stimuli. The Bibliography provides additional references to such experimental work.

The Protozoa represent a group of organisms whose basic characteristic is that they are acellular. They possess no organized nervous system and yet they are able to respond to certain stimuli. The responses they make are limited and relatively easy to correlate with a particular stimulus. However, rigorous study of some aspects of their behaviour has shown that not all their responses are quite as simple as they appear at first sight.

E3.1 Investigation of behaviour in Protozoa (*Spirostomum* sp.)

Spirostomum ambiguum is an acellular freshwater protozoan belonging to the Ciliates, as its method of movement is by rows of cilia covering the body. It is found in ponds, particularly those with a high content of organic material. It is a convenient protozoan to use, as it is much larger than most. Alternatively, *Paramecium* could be used, although this is a much smaller organism.

A The normal movements of *Spirostomum*

Procedure

1   Take a microscope slide, add a few drops of a culture containing some *Spirostomum*, and examine them; use either a low power binocular microscope or a microscope with a ×4 objective. It may be found easiest to arrange the slide on a black surface with the light falling on it from the side.

2   Observe the shape of the animal and compare it with the illustration (figure 61).

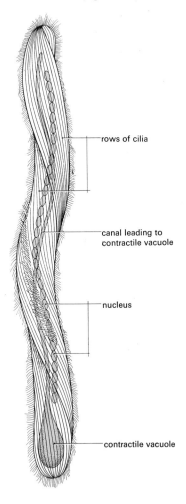

rows of cilia

canal leading to
contractile vacuole

nucleus

contractile vacuole

Figure 61
Diagram of *Spirostomum abiguum*.
*From Mackinnon, D. L. and Howes,
R. S. J. (1961)* Introduction to the
study of Protozoa, *Oxford University
Press.*

3   Follow carefully the locomotion of the animal, noting any peculiarities in orientation as it proceeds.

4   Tap the slide with a pencil, while still observing the animal, until a definite effect is shown.

5   Add some very small pieces of soaked filter paper to form a barrier in the centre of the slide and study carefully the behaviour of the animal when it reaches this barrier.

6   Pour some of the liquid from the *Spirostomum* culture vessel into a container such as a crystallizing basin. Allow the debris to settle and observe the movement of the animals; a stereo binocular microscope may help in this.

Questions

a   How did the animal react when the slide was tapped?

b   Does the movement of the animal after striking a barrier bear any relation to the previous direction of movement or to the position of the barrier?

c   Are the movements of the protozoans in a large container in any way different from those observed on a slide?

Having studied the normal reactions of the animal, you can alter its environment by one factor at a time and observe the effect. The best way to record the results may be in the form of a table with space provided for comments. If there is uncertainty about the response in a particular situation, try to repeat the experiment; but in any event, compare your results with those of other students. Remember that it is possible for some animals to give different results in seemingly identical conditions. It is important in all these experiments to use liquid without any debris in it, as the organisms may tend to remain in this area and not respond to other stimuli.

B   Response of *Spirostomum* to gravity

Procedure

1   Take two U-shaped pieces of glass tubing, the arms of which should be about 5 cm long and external diameter 6–7 mm (bore 4–5 mm).

2   With a glass marking pencil divide each tube into 5 sections of equal length, one along the base, and two up the sides, the last section finishing about 1 cm from the top.

3   Fill each tube to the level of the top section with culture, using a dropper pipette containing about 10–15 *Spirostomum*.

4   Set the tubes upright; a piece of Plasticine makes an effective base.

5   Leave one tube exposed to even, diffuse illumination from the side (daylight from a window would be suitable), but not direct sunlight.

6   Carefully wrap the other tube in aluminium foil so as to exclude all light.

7   Examine the tubes every 20 minutes, disturbing them as little as possible.

8 Record the number of the animals in each of the sections. If there has been no detectable reaction by the end of the practical session leave the tubes overnight and record the results the following morning.

Questions

a Do the animals show a reaction to gravity? A positive reaction would be indicated if they tended to collect at the bottom, a negative one if they collected at the top.
b How long did it take for a definite response to occur (assuming that there was one) and did all the individuals show the same response?
c Was there any detectable difference between the animals' response in the tube in the light and that kept in the dark? If there was a difference, can you suggest hypothesis to account for it?
d In this experiment we have tried to control the light, but are there any other factors which might have led to differential distribution? If so, could you design an experiment to test their effect?
e In what ways would you criticize the design of the experiment? Can you think of a better procedure?

C Response of *Spirostomum* to light

Procedure

1 Place a microscope slide on a black background and add enough *Spirostomum* culture to form an elongated 'pool' about 4 cm long and 1 cm wide containing about 10–20 animals.
2 Illuminate the pool evenly with a bench lamp and watch the animals for 1–2 minutes. See if they appear to be evenly distributed or if there is any tendency to collect in one part of the pool. A hand lens will be sufficient for this.
3 Now take a piece of aluminium foil and bend this to form an arch. Fit this over one end of the pool to shade half of it from the light.
4 After 10 minutes observe the distribution of the animals and record the number in the shaded and non-shaded portions. Repeat this after a further five minutes or until you have obtained a definite result.
5 Now transfer the cover to the other end of the pool and repeat the observations.

Questions

a Was the initial distribution in the pool even? If not, can you suggest what might have caused an uneven distribution?
b Do your results indicate that the animals react positively or negatively to light?
c Did all the animals react in the same way? If not, can you suggest any explanation for the apparent difference in behaviour?

d What was the purpose of transferring the cover to the other end of the pool? Was the second result similar to the first?

e What adaptive value do you think any response that you have detected may have?

D Response of *Spirostomum* to chemicals

Procedure

1 Again make a pool of *Spirostomum* culture on the slide against a black background. Repeat this with a second slide and set up both so that they are illuminated with even, diffuse light.

2 Now take two strips of filter paper, one of which has been soaked in sodium chloride solution and the other in a weak acid. Make sure that you keep the pieces separate.

3 From each strip cut about 10 rectangles each 2 mm by 4mm.

4 Place two of the rectangles which have been soaked in sodium chloride solution, one at a time, on the extreme left-hand edge of the pool on one slide. The easiest way to do this is to pick the piece up with a pair of forceps, place it on the slide, and push gently up to the edge of the pool.

5 Repeat this procedure, but using pieces of acid paper, with the other slide.

6 Observe closely the movement of the animals towards or away from the end with the pieces of filter paper. If no movement is visible, after 2–3 minutes add a further piece of the appropriate filter paper. Continue this procedure until a definite response has been obtained.

7 Record the result together with the number of pieces of filter paper added.

Questions

a Why was it necessary to illuminate with even, diffuse light?

b What was the advantage, if any, of adding the pieces of filter paper one at a time, rather than all together?

c Did you observe any unusual reactions of *Spirostomum* during this experiment?

d How many pieces of salt paper and acid paper did you finally add?

e How could you be reasonably certain that it was the chemicals and not the filter paper that caused the animals to respond?

f What biological advantage might the responses you have observed give the animals in their natural habitats?

E3.2 Behaviour of molluscs

The common garden snail and slug are easily obtainable organisms and can be used in simple experiments to investigate movement in response to various stimuli. Again you will have to design your experiment carefully to eliminate, as far as possible, all stimuli except the one you are investigating.

Procedure

1   The animals leave a trail of slime behind them when they move, and this can be utilized to record their movement. Take a clean, dry, polished sheet of glass or Perspex; 25×20 cm is a convenient size. With a glass marking pencil number the four corners.

2   Clamp the glass in the desired position and release the organism in the centre. The animals vary in their activity so you may have to select the more active ones. Activity is enhanced by warm conditions.

3   When the animal has reached the side of the glass move it, using forceps, to a labelled container. Dust the glass with talc and tap to remove the excess. The 'mollusc' trail will show up as a white line. This can be recorded on quarto paper (for a 25×20 cm sheet of glass) by tracing or copying. Remember to number the corners of the sheet of paper.

4   The experiment must be repeated with a number of the same species of mollusc and the animals must be kept in separately labelled containers.

Questions

a   Do molluscs show a response to gravity? Does the movement of the animals on a vertical surface differ from the movement made by the same animals on a horizontal surface?

b   What is the effect of repeating the experiment using a vertical surface submerged in water? Instead of blowing the excess talc away on the wet plate, gently wash it away.

c   Can the animals detect the presence of food material placed at the edge of the glass sheet? Do they show any preference for different food materials?

d   Do the animals respond to a light beam?

Courtship behaviour in birds

Birds have proved useful organisms for the investigation of courtship and parental care. You may already have some knowledge of the behaviour of terns and gulls.

Some species of cage bird, such as the zebra finch, *Taeniopygia castanotis*, will readily court in captivity.

E3.3 Courtship behaviour in the zebra finch (*Taeniopygia castanotis*)

The zebra finch is a member of a family of Australian grass finches. The animal has been domesticated in this country for some years and is a common cage bird.

Courtship by the male

Observe the two sexes and note the sexual dimorphism. As is common in most birds, the male possesses the more distinctive plumage. At the start of the investigation the sexes should have been separated for some time. It is best if individual birds have been marked with coloured rings on their legs so that they can be easily identified.

There are three main components of the male finch's courtship, although there may be wide variations in the activity and intensity with which they are expressed.

1 Song–During courtship the male repeatedly sings a short song phrase. This may change to a loud, rapid single call.
2 Posture–The posture of the male responding to a female is very characteristic. Figure 62 illustrates this.

Figure 62
The courtship posture of the male zebra finch.

3 Dance–As the male advances towards the female he usually performs a pivoting dance, as illustrated in Figure 63. The various components of courtship occur together and you may find it difficult, at first, to analyse it satisfactorily.

Procedure

1   You may find it useful first to watch a film of the birds' behaviour. The intensity of the various responses differs considerably between individuals and at different times. Behaviour is also affected by environmental disturbances, so you should ensure that the room is kept quiet and you should move about as little as possible.

Figure 63
The courtship dance of the zebra finch.
From Morris, D. (1954) Reproductive behaviour of the zebra finch (Poephila guttata) Taeniopygia castanotis *with special reference to pseudofemale behaviour and displacement activities,* Behaviour, **6,** *pp. 271–321.*

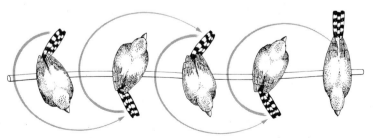

2   Introduce a female into an empty observation cage and after she has settled down release a male bird with her. Observe the reactions of both birds for the next two or three minutes and record their behaviour.

3   After three minutes separate the birds again.

4   The male bird must be able to recognize the female. She is rather a dull colour, but one outstanding feature is the red beak, which has been interpreted as the sign stimulus eliciting courtship in the male. It is believed to play an important part in recognition.

5   You can investigate this by using a model, constructed from plaster of Paris or white Plasticine. It does not have to resemble a female very closely. The beak can be coloured red with nail varnish. Place the model on a perch in an empty cage and then introduce a male. Observe and record his behaviour during the next two or three minutes.

6   A further investigation could be to alter the colour of the female's beak to that of a juvenile. To do this carefully catch a female finch and holding her gently but firmly, colour the beak, using a black felt marker pen. Release her into an empty cage and when she has settled down introduce a male bird.

7   Carefully record the responses of the male over the next two or three minutes. If possible, repeat this with other combinations of birds.

Questions

a   What characteristics of the male bird distinguish it from the female?

b   In what ways does the courtship posture in the male differ from the normal stance?

c   What parts of the male are displayed in the courtship postures?

d   What responses did the females make towards the male? Describe these in terms of (1) auditory responses, (2) posture, and (3) movement.

e   Apart from the reactions of the male described above did you observe any other behaviour on his part?

f   If the male courted the female, how long did this activity last?

g   What sort of variation in behaviour did you find in different birds?

h   How did the male respond to the model and to the black beaked female?

i   Observation of courtship in wild birds is possible even in urban areas. The sparrow and pigeon are suitable inhabitants of such localities. Consult the Bibliography for descriptions of behaviour. What responses can you identify in the birds of your area?

Bibliography

Carthy J.D. (1958) *An introduction to the behaviour of invertebrates*. Allen & Unwin. (Much detail of work on the orientation of invertebrates.)

Carthy, J.D. (1966) *The study of behaviour*. Edward Arnold. (Chapters 2 and 3 on orientation and courtship.)

Evans S.M. (1969) 'The study of bird behaviour in schools.' *Journal of Biological Education*, **2**, 373-380. (Use of zebra finches with ideas for futher experimental work.)

Fraenkel, G.S. and Gunn, D.L. (1961) (1940) *The orientation of animals*. Dover Press. (The best known book on the subject, which gives much detail on the classical experiments, many of which are suitable for schools.)

Goodwin, D. (1961) *Instructions to young ornithologists, Volume 2: Bird behaviour*. (Simple account of some aspects of their behaviour.)

Meyerrieks, A.J. (1962) *Courtship in animals*. B.S.C.S. Pamphlet No. 3. D. C. Heath. (Survey among the animal kingdom.)

Murton, R.K. (1965) *The wood pigeon*. Collins. (Contains much information applicable to the feral urban pigeon.)

Snow, D.W. (1958) *A study of blackbirds*. Allen & Unwin. (Details of behaviour of another common species.)

Summers-Smith, J.D. (1963) *The house sparrow*. Collins. (Courtship behaviour.)

Tinbergen, N. (1953) *Social behaviour in animals*. Methuen. (Courtship and mating behaviour.)

INDEX

stimuli, 51–5
  in courtship behaviour, 103, 105
  reactions to, 56–67
stomata of leaves, and loss of water, 6
strychnine, poisoning by, 77
sucrose, plasmolysis of plant cells by
    solutions of, 22
sulphur, in plant nutrition, 45
summation, of stimuli to muscle, 77
survival, behaviour and, 52
swallows, Gilbert White on, 52
swans, peck order among, 102
sweat, loss of water in, 26

t

Taeniopygia castanotis (zebra finch),
    courtship of, 120–2
taxis, 54
temperature
  and development of larvae in
      groups, 100
  and life cycle of *Tribolium* spp., 12
  and uptake of water by plant
      tissues, 42–3

territories, claims to, 53, 107
tetany, 77, 78
Tinbergen, N., 103
transpiration stream of plants, 2, 3, 6
Tribolium (flour beetle)
  characteristics of two spp. of, 12
  responses of, to humidity, 12–14, 54
Triticum (wheat), coleoptile of, 59
tropisms, 59
turgor, 20–1, 23–4

u

ureter, 29
urinary system, 27–9
urine, 29, 34
  composition of, 39
  hyperosmotic, 36
  loss of water in, 26
Urtica dioica (nettle), optimum water
    content of soil for, 2

v

visual signals, and social behaviour,
    107, 108
Volta, A., 72

w

water
  animals and, 10–14
  balance of, in mammals, 25–40
  the cell and, 16–24
  plants and, 2–10
  and survival, 1–2
water table, 2
White, Gilbert, 52
Whytt, Robert, 80
wilting of plants, 1, 23
woodlice, responses of, to
    environmental stimuli, 11, 97

x

xylem, 9, 10

z

zinc, in plant nutrition, 46